I0422262

ISBN: 9798884873636

Cover design by: Art Painter
Library of Congress Control Number: 2018675309
Printed in the United States of America

LIVE LONGER
LOVE LONGER
Age is a treatable disease.

G SYED MD FIPP

TABLE OF CONTENT

References
Word List: Tricky Book Terms

Dedicated to the Umar Syed Foundation,

Your mission is so noble, that even Santa Claus asked for a grant application form. Seriously though, your dedication to education and health for indigent communities is like the superhero of nonprofits – saving the day, one stethoscope and textbook at a time.

We're donating the earnings from "Book Live Longer and Love Longer" because, let's face it, laughter is the best medicine, but education and healthcare come pretty close. So here's to helping families and kids suffering from mental illnesses with addiction – may they find the strength to laugh a little louder and love a little longer.

Keep doing what you do best – making the world a better place, one belly laugh and prescription refill at a time.

With giggles and gratitude,Love Longevity: Treatable Aging

"Extend Your Existence, Extend Your Romance:
Chronicling Aging as a Curable Malady"
For Grown-Up Kids

INTRODUCTION

Love Longevity: Treatable Aging

Ah, the eternal quest to cheat time and stick around this crazy world a little longer!

Picture this: you're chilling in a lab with Dr. Murli Pagal, surrounded by mice that seem to have found the Fountain of Youth. We're talking about tweaking aging in these little rodents and dreaming of doing the same for us humans.

Flashback to May 14, 1997: I'm at the New York Academy of Medicine, presenting a paper that's a love letter to longevity. We're diving deep into the world of aging mice, contemplating whether we can pull off the ultimate heist against Father Time.

Fast forward a few decades, and here we are, still pondering the mysteries of aging like a bunch of scientific sleuths. But hey, we've made progress! We're not just talking about living longer; we're talking about loving longer too. Aging isn't just a fact of life anymore; it's a treatable condition!

Enter "Love Longevity: Treatable Aging," the book that's shaking up the status quo and asking the big questions. Can we kick aging to the curb? Can we flip the script on decay and degeneration? Can we party like it's 2099 well into our golden years?

This book isn't just a bunch of lofty ideas; it's a roadmap to hacking your lifespan. From cutting-edge science to practical tips, it's all here. Say goodbye to just existing and hello to thriving well past the century mark. So if you've ever pondered the secrets of aging or dreamt of hitting triple digits with style, dive into "Love Longevity" and prepare to have your mind blown!

Alright, folks, gather 'round for the wisdom of the ages, or should I say, the wisdom of four decades of dabbling in functional medicine! Strap in, because we're about to embark on a journey to unlock the secrets of eternal youth—or at least, to make aging less of a pain in the neck.

Picture this: you're lounging on your rocking chair, flipping through the pages of "Live longer, love longer, age is a treatable disease." Yep, you heard that right, we're treating aging like it's the common cold. Who needs wrinkles when you can have vitality for days?

But wait, there's more! This book isn't just about adding more candles to your birthday cake; it's about increasing the quality of your health span. We're talking about living so long, you'll need a GPS to find your birth certificate. And with all the fancy-schmancy advancements in scientific medicine, hitting the triple-digit milestone is practically a given.

So, if you're tired of feeling like you're running on expired batteries, this book is your ticket to the fountain of youth. Learn the secrets of outsmarting aging, from tweaking your diet to mastering genetic therapy. Imagine a world where you're not just hanging out with your grandkids but swapping stories with your great-grandkids.

Dr. Syed here isn't just spouting off pipe dreams; he's laying down a roadmap to a future where we're all sipping on the elixir of everlasting life. So, grab your walking cane and join the revolution—because we're about to rewrite the book on aging, one laugh line at a time!

Ah, behold the holy grail of anti-aging wisdom: "Live longer, love longer, age is a treatable disease." Because who needs wrinkles when you can have a VIP pass to eternal youth, am I right?

This book isn't just a stroll down Memory Lane; it's a sprint toward the Fountain of Youth, fueled by groundbreaking science and enough optimism to power a small country. We're talking about transcending the very fabric of time itself to snag a longer, healthier, and happier existence. Forget aging gracefully; we're aging disgracefully, in the best possible way!

And hey, if you're tired of feeling like yesterday's news, fear not! This book doubles as a roadmap to the Promised Land of eternal youth. We're talking about merging scientific wizardry with lifestyle hacks so simple that even your grandma could do them. It's like turning back the clock so far, you'll be carded at the retirement home.

But wait, there's more! Aging isn't just a fact of life; it's a bonafide disease, complete with its own fancy ICD-11 code: MG2a - Old Age. Why settle for regular old age when you can have personalized medicine catered to your every wrinkle and gray hair?

And don't even get me started on the treatable hallmarks of aging. We're talking about genomic instability, telomeres shorter than your attention span, and epigenome shenanigans that make your genes want to party like it's 1999. But fear not, dear reader, because we've got the antidote: demethylation, deacetylation, and a sprinkle of good old-fashioned genetic tinkering.

So buckle up, buttercup, because we're rewriting the model of life and death faster than you can say "Benjamin Button." Who needs a mere mortal lifespan when you can live like a Caenorhabditis Elegans worm on steroids? Here's to a future where age is just a number and wrinkles are nothing but a distant memory!

Ah, behold the wonders of emerging technologies! Forget stem cells, folks, because we've got something even better: molecule reversine! It's like hitting the rewind button on your cells, turning those tired old wrinkles into fresh, youthful skin faster than you can say "Benjamin Button's skincare routine."

Sure, your birth certificate might stubbornly insist on your chronological age, but who cares when you can reverse your biological age? It's like swapping out your old, beat-up jalopy for a shiny, new sports car—except in this case, the sports car is your body, and the upgrades are molecular magic.

But wait, there's more! Our country's health goals are about to get a serious makeover. We've already upped the ante on lifespan, but now it's time to tackle the real challenge: health span. Because what good is living longer if you're as miserable as a clown at a funeral?

And speaking of misery, let's talk about mental and emotional health. Sure, we've conquered smoking and infectious diseases, but have we tackled the big bad wolf of loneliness and anxiety? Turns out, social connection is the secret sauce to a longer, happier life. Who knew that ditching your hermit lifestyle could be the key to dodging depression and dementia?

But fear not, dear reader, for the power to age like a fine wine, is within your grasp. It's as simple as breaking a sweat, munching on some kale, and not avoiding the doctor like they're the plague.

With a little help from science and a lot of help from common sense, you too can defy the odds and live your best, most ridiculous life well into your golden years.

Cheers to aging disgracefully, my grown-up kids and senior friends!

Preface

Forever Young: Cracking Age Secret

Once upon a time, in the land of humans, we've been around for ages, but our bodies seem to have a funny way of showing it. We're living longer than ever, but sometimes it feels like we're just sticking around, not enjoying life much.

Picture this: rich folks in fancy countries often spend years feeling yucky with one sickness after another before they finally say goodbye. And guess what? This might become the norm for lots more people as time goes on.

Doctors say our fancy medical treatments are making death more of a medical affair. So instead of going peacefully in our sleep like Grandma's stories, we might end up hooked to machines and swallowing bitter pills for ages. Yuck! And don't even get me started on the bills. Hospitals must think we're made of money!

But hey, what if we could change the game? What if we could stay sprightly and lively for way longer? Imagine being as playful as a kid even when you're as old as a grandpa! Sounds wild, right?

What if our last years were as awesome as our younger ones? What if we could skip the stress of growing up too fast and take our time enjoying life's adventures?

Imagine hitting middle age and feeling as fresh as a daisy, ready to try new things without worrying about the clock ticking away. And what if I told you, this might happen sooner than you think?

For ages, I've been on a quest to figure out why we get old. It's been like trying to find the beginning of a river – tricky stuff! But after lots of twists and turns, I might have stumbled upon the secret sauce.

Alright kiddos, buckle up for some mind-bending stuff! In the next chapters, I'm about to drop some knowledge bombs about why we get old and what we can do about it.

First off, in section 1, I'll dive into the past and explain why I think aging is the ultimate sneaky disease. Yep, you heard me right – getting old is a big old problem that needs fixing!

Then, in section II, I'll talk about what's happening right now to tackle this aging business. I'll spill the beans on all the cool tricks and treatments scientists are cooking up to slow down, stop, or even reverse aging. Imagine never having to worry about getting old – sounds like a dream, right?

And finally, in the last section all about the future, I'll lay out some wild ideas about what could happen if we put a stop to aging as we know it. Spoiler alert: it's going to be epic! We're talking about a world where getting older means staying healthier for longer – no more achy joints or wrinkly skin, just living our best lives without a care in the world. Sounds pretty awesome, huh? So, get ready to join me on this journey to a future where we're all living our longest, healthiest lives ever!

Alright kiddos, listen up because we're about to dive into some serious doubting Thomas territory. You know, those folks who think aging gracefully is as likely as finding a unicorn in your backyard.

They'll tell you all sorts of gloomy stuff like how our modern ways are like a curse, shortening our lives quicker than you can say "snail on roller skates." And forget about hitting the big 1-0-0; they reckon it's as rare as finding a four-leaf clover in a haystack. And if by some miracle you do make it to a century, well, brace yourself for a not-so-fun ride filled with aches, pains, and more doctors' visits than you can shake a stick at.

They'll throw some fancy numbers at you, talking about how we've been inching our way up the life expectancy ladder since the dawn of time. But here's the kicker: while the average lifespan keeps creeping up, there's a pesky little limit that doesn't budge much. It's like trying to stretch a rubber band beyond its breaking point – not gonna happen easily, folks!

But hold on to your hats because here's where things get spicy: some smart cookies out there think we can do better. Yep, you heard me right – they reckon we can not only tack on more years to our lives but also keep those years full of pep, vim, and vigor. Imagine skipping through life like a spring chicken well into your golden years – sounds like a blast, huh?

And get this: they're not just blowing hot air. Nope, they've got the science to back it up. They say there's no biological rulebook saying we gotta get all creaky and wrinkly as the years roll by. They reckon we're on the brink of a whole new era where living to a ripe old age is the norm, not the exception.

So buckle up, buttercups, cause we're on the brink of a revolution

– an evolution, even! It's time to rewrite the rulebook on what it means to be human and say goodbye to the old ways of aging. Who's ready to join the party?

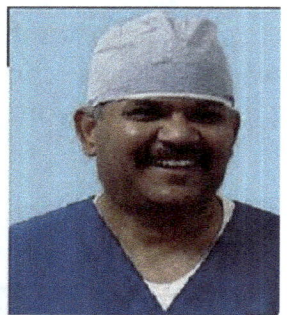

G SYED MD FIPP

CHAPTER 1

Life Origins: Life Soup

THE BEGINNING: How Life Started:Let's take a journey back in time to when life first began on our planet.

Picture a world that's kind of like ours but with some big differences. It's smaller, spins faster, and has a strange mix of gases in the air, but no oxygen at all. There are no plants or animals, just some islands poking out of a salty ocean.

This world is tough! It's hot, volcanic, and wild. But in one spot, near some hot underwater vents, something magical is happening. Organic molecules are gathered and brought in by comets and meteorites. When these molecules mix with warm water, they start to change. They form long chains called polymers, kind of like when salt crystals form on the beach.

Then, when the water evaporates, these molecules get enclosed by fatty acids, creating tiny bubbles - the first cell membranes! Over time, these bubbles become filled with genetic material, like the ingredients for life soup.

As the pools fill up again, these bubbles multiply, creating a frothy layer on the water's surface. It's like a microscopic soap opera as they compete to survive. Some evolve into more complex forms and spread into rivers and lakes.

But then, a big challenge arises - a long dry season. The lakes dry up completely, leaving behind a thick, yellow crust. It's a tough time for these early life forms. They must fight hard to survive because they're the ones who will kickstart all life on Earth - from tiny bacteria to big animals like us!

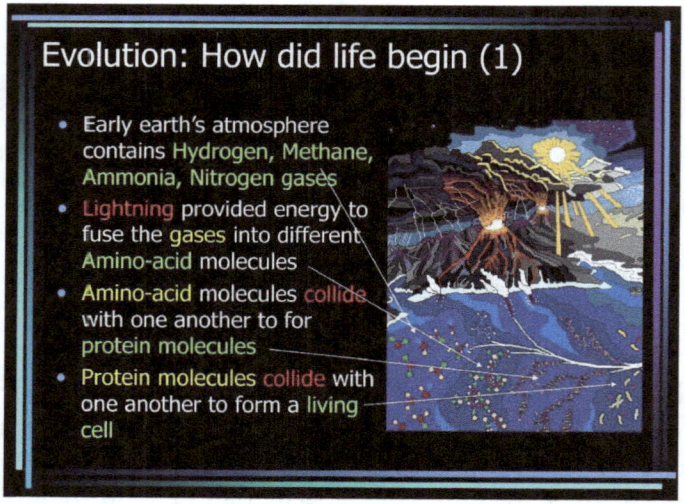

Evolution: How did life begin (2)

- At first, life was very simple and consist of "self replicating molecules"
- Later, life learn to store genetic information in DNA molecules: single cell organisms (bacteria)
- And later, single cell organisms developed into simple multi-cell orgasms (Example: coral polyp)

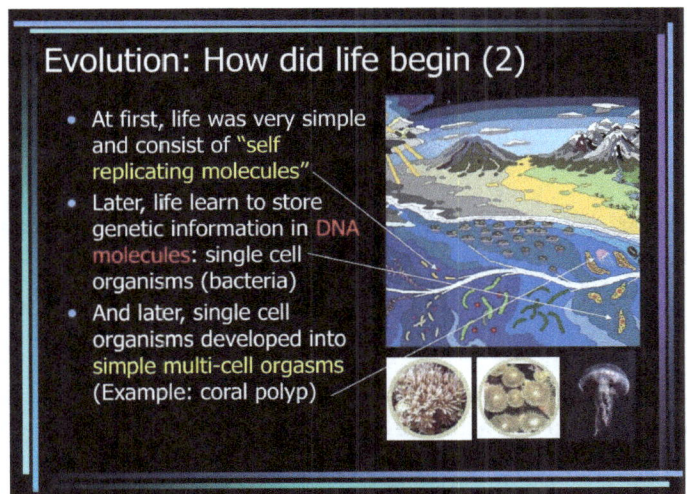

Amid this chaotic cell party, where everyone's fighting for scraps of food and drops of water, there's this special dude called Magna superstes, or "super survivor" in fancy talk. Now, Magna superstes might not look like much, but it's got a secret weapon: a super cool genetic trick.

Okay, fast forward billions of years and life gets crazy complicated. We're talking about creatures with eyes, brains, and all sorts of fancy stuff. But back in the day, Magna superstes kept it simple with a little gene circuit.

Picture this: there's gene A, the cell's responsible babysitter, making sure it doesn't go crazy and multiply like crazy when times get tough. Then there's gene B, which codes for a protein that tells gene A to chill out when everything's cool so the cell can have baby cells without worry.

Now, here's where Magna superstes gets its superhero cape. Gene B has a mutation that turns it into a DNA repair wizard. So, when the cell's DNA is all messed up, this protein rushes to the rescue, fixing things up like a DNA handyman. And while the DNA's getting patched up, the cell hits pause on the whole baby-making thing because, let's face it, you don't want to pass on messed-up DNA.

Once the repairs are done, Magna superstes springs back to life, ready to party and make more copies of itself. It's like hitting snooze on reproduction until the DNA's back in tip-top shape. And that's how Magna superstes becomes the MVP of survival, spreading its genes and creating generations of new descendants like a boss. They're basically the OGs of life on Earth.

Like the legends of Adam and Eve, we're not sure if Magna superstes ever really existed. But research over the past twenty-five years suggests that everything alive today owes a little something to this super survivor, or at least to something pretty similar.

Check it out: every living thing on this planet, from plants to animals to us humans, carries around this ancient genetic survival kit. It's like a timeless recipe for staying alive in a world that can be both harsh and generous at times.

This survival kit is so awesome because it knows how to prioritize. When things get tough and our genes start freaking out, it directs all our energy to fix the problem, making sure we're in tip-top shape before even thinking about making baby cells.

And this survival kit has been passed down through generations, getting better and better over billions of years. It's like the ultimate hand-me-down that keeps on giving, helping us survive even when the universe throws us curveballs.

Now, here's the kicker: this survival kit is also the reason why we age. Yup, you heard that right. As awesome as it is, it comes with a downside. But hey, it's a small price to pay for being part of the lineage of great survivors, right? We might age, but we're still kicking and thriving, proving that we're the real MVPs of evolution!

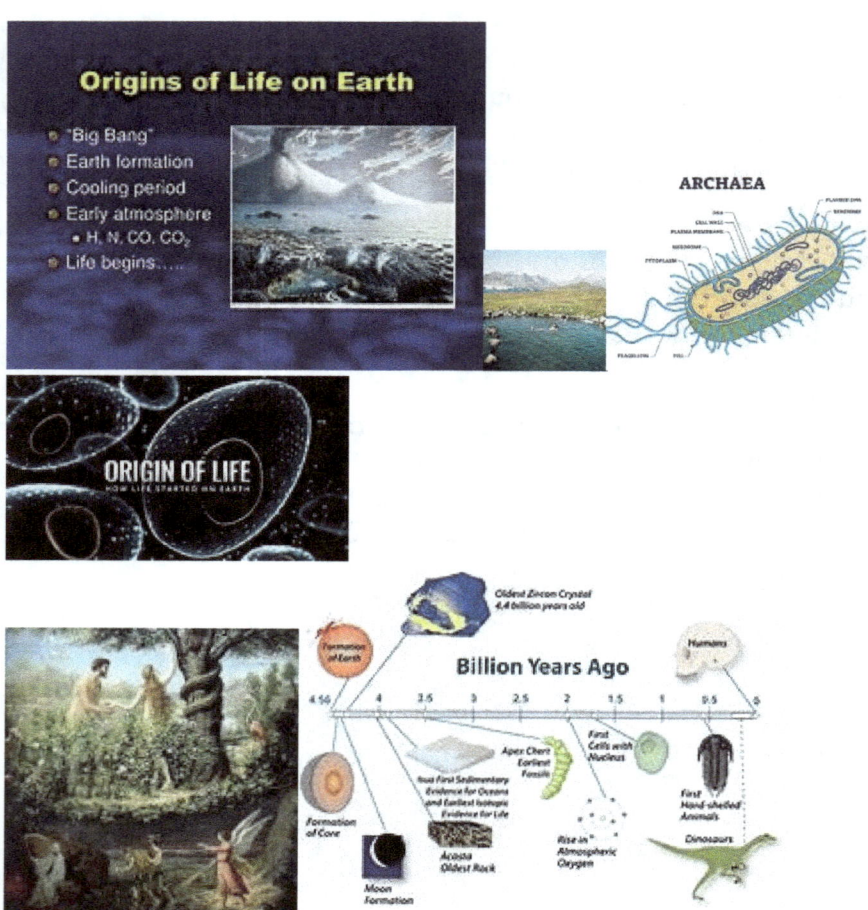

Okay, picture this: way back when life was just starting, billions of years ago, there was this super ancient gene circuit that had a nifty trick.

Gene A was like the boss, telling cells to hold off on making baby cells when things get dicey. And then there's gene B, which made a protein that told gene A to chill out when everything was cool for some baby-making action.

But here's where it gets wild: when the DNA got messed up, the protein made by gene B left its baby-making duty and went to fix the DNA. And while it was busy with repairs, gene A kicked in to stop any baby-making until everything was back to normal.

Now, fast forward to today, and we've got a souped-up version of this survival circuit keeping us going strong. It's like inheriting the ultimate DIY repair kit from our ancient ancestors, ensuring we keep ticking even when life throws us a curveball.

I'm unable to create diagrams directly in this text-based format. However, you can easily sketch a diagram illustrating the gene circuit described. Here's a simple description of how you could draw it:

1. **Gene A**: Draw a circle labeled "Gene A" and inside it writes "Halts reproduction when DNA breaks."

2. **Gene B**: Draw another circle labeled "Gene B" and inside it writes "Make protein to silence Gene A during normal times and repair DNA when damaged."

3. **Protein**: Draw an arrow from Gene B to represent the protein it produces. Label the arrow "Protein" and indicate that it moves to repair DNA when needed.

4. **Action during DNA repair**: Draw another arrow from Gene B's to Gene A's, indicating that when the protein leaves to repair DNA, Gene A becomes active, halting reproduction until repairs are complete.

You can draw these elements on a piece of paper or using any digital drawing tool you prefer to create a visual representation of the gene circuit.

◆ ◆ ◆

Discovering the Aging Cause

Why We Get Old - because everything has a reason;

Okay, so if you're like, "Wait, why do we even age?" You're not alone. Loads of folks haven't really thought about it. Even some fancy biologists and aging experts are like, "Eh, who knows?" They're just busy dealing with the fallout of getting old.

It's kind a like how back in the day, people were clueless about cancer. They just tried to hack away at tumors and told folks to get their stuff together. Cancer was just this mysterious thing nobody could explain.

Then in the '70s, some smarty-pants scientists found out about these genes that go haywire and cause cancer. Suddenly, everything changed. Now we could target those pesky proteins causing trouble and zap 'em without hurting the good cells.

We're not curing all cancers yet, but we're way more hopeful. Like, there's this cool thing called immunotherapy. Our immune system is like a superhero, always on the lookout for bad cells. But cancer cells can be sneaky and hide. Immunotherapy basically takes off their invisibility cloak so our immune system can kick their butts.

Right now, only a small group of folks benefit from it, but there are tons of trials happening, so hopefully, more people will get the help they need soon.

The Drama of Getting Old -
It's a Rollercoaster Ride

Okay, so picture this: getting old is like solving a super tricky puzzle. We've been scratching our heads, trying to figure out why we age. But it's no walk in the park. We need a solid explanation that covers everything from physics to chemistry to biology. Like, it's gotta be epic, spanning from tiny molecules to big living things.

But guess what? There's no one-size-fits-all theory of aging. People have thrown out ideas, like DNA getting messed up or free radicals running wild, but none of them really stuck.

Remember that dude Szilard? He was all about trying to clone humans, which turned out to be a game-changer. If aging was all about losing genetic info, cloned animals would be born old, right? But turns out, they're not! Cloned animals live just as long as regular ones, showing that mutations aren't the main aging culprit.

So, basically, science is like a rollercoaster ride. We're in this crisis mode where old theories are getting tossed out the window. But hey, that's how progress works, right? We're on the brink of a new era where we'll finally crack the aging code. Hang on tight, folks!

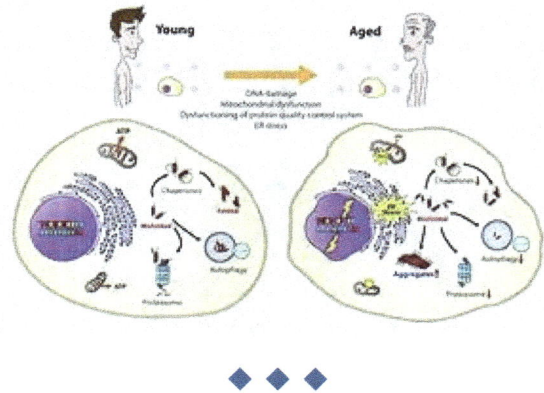

The 9 Big Signs of Getting Old:

1. **DNA Mix-Ups:** Picture your DNA like a recipe book. Sometimes, stuff gets spilled, pages tear, and recipes get messed up. That's DNA damage for you.

2. **Shortening Telomeres:** Imagine your chromosomes like shoelaces with little caps at the ends. As you age, those caps wear down, causing trouble.

3. **Epigenetic Changes:** It's like a switchboard controlling which genes are active or not. As you get older, someone's messing with the switches.

4. **Protein Maintenance Fail:** Think of proteins like your body's workers. As you age, some slack off or get fired, leaving a mess behind.

5. **Wonky Metabolism:** Your body's metabolism is like a factory. With age, the machines start acting up, causing chaos.

6. **Mitochondrial Mishaps:** Mitochondria are the powerhouses of your cells. As you age, they start slacking, leaving you feeling tired and sluggish.

7. **Zombie Cells:** These are cells that should've retired but keep hanging around, causing trouble for the good cells nearby.

8. **Tired Stem Cells:** Stem cells are like the superheroes of your body, but as you age, they start feeling exhausted and can't do their job properly.

9. **Bad Communication:** It's like your cells stop talking to each other properly, leading to misunderstandings and chaos.

Researchers think if we tackle these signs head-on, we can slow down aging, prevent diseases, and maybe even delay death.

For example, if we can keep stem cells feeling fresh, they can keep fixing up our bodies and fighting diseases. And by getting rid of those pesky zombie cells, we can keep our tissues healthier for longer.

Scientists are working hard on solutions for each of these signs, using fancy tech and tons of data crunching. But we're still in the early days of figuring it all out. It's like we've cracked the code, but there's still a lot of mystery to solve.

But hey, with all this brainpower and technology, who knows? Maybe we'll finally unlock the secrets to living longer and healthier lives. It's an exciting time to be alive!

nine
Hallmarks of Aging

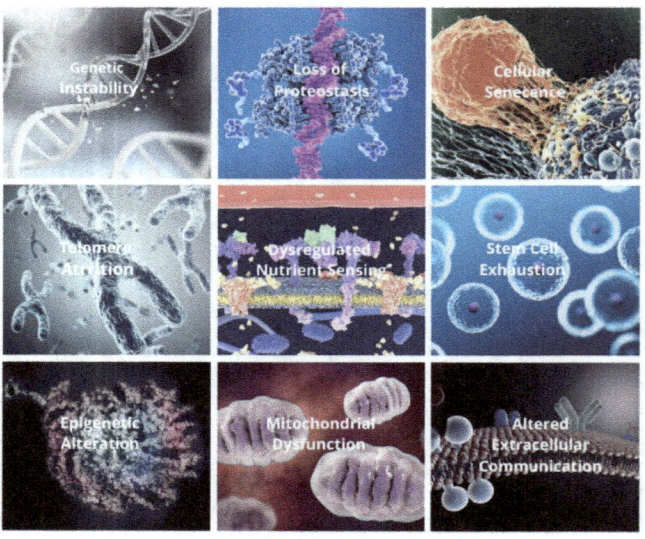

Getting Old: Losing Info.

L osing the Instructions.

Okay, so imagine if there was this big secret behind why we get old. It's like trying to solve a really tricky puzzle, but way harder because it's about getting old, which is super complicated. Scientists have been scratching their heads over this for ages.

Now, to understand this puzzle, let's talk about two types of information in our bodies. First, there's digital info, kind of like the stuff in a computer. It's all about DNA, those tiny codes that decide everything about us. DNA is like a super durable storage system, keeping our info safe and sound for billions of years.

Then, there's the second type: analog info. It's not as famous as DNA, but it's super important too. Analog info is all about how our genes behave and switch on and off. It's like the software that tells our cells what to do.

So, while DNA is like the blueprint of a house, analog info is like the instructions telling the builders how to put it all together.

Now, imagine our body is like a DVD player. Over time, it starts getting a bit glitchy, like when your favorite DVD skips or freezes. That's kinda what happens to us when we age. But here's the cool part: even though our bodies get old and glitchy, our DNA stays young at heart. It's like having a shiny new DVD hidden inside that old, scratched one.

So, if we can figure out how to polish away those scratches and glitches, we might just unlock the secret to staying young. It's like fixing up that old DVD player to play your favorite movie again.

And hey, if scientists can figure this out, we might just be able to turn back the clock and stay forever young!

The Great Aging Adventure!

Okay, picture this: our bodies are like a big adventure playground, full of secret codes and tiny helpers that keep us running smoothly. But as we grow older, some of these codes start to fade, like when your favorite game gets a little glitchy.

But fear not, because scientists have uncovered some super cool secrets about how our bodies work! It's like discovering hidden treasure in a treasure map!

First up, we've got these genes called "longevity genes." They're like the guardians of our bodies, helping us stay strong and healthy. They're kind of like superheroes inside us, fighting off bad stuff that tries to make us sick.

One group of these genes is called "Sirtuins." They're like little repair crews, fixing up any damage in our cells and making sure everything runs smoothly. And guess what? They love a molecule called NAD, which gives them their superpowers.

Then, there's another group called "TOR." They're like the traffic controllers in our bodies, telling cells when to grow and when to chill out. And when things get tough, they kick into action, helping us stay strong and resilient.

Last but not least, there's "AMPK." They're like the energy police, making sure we've got enough fuel to keep going. And when energy levels drop, they step in to save the day!

So, next time you feel like your body is getting a little glitchy, just remember you've got a whole team of superheroes inside you, ready to keep you going on your amazing adventure called life!

Alrighty, let's dive into a world of tiny superheroes inside our bodies called "Hormesis"! These little guys are like the guardians of our health, and they love a good challenge.

When our bodies face a bit of stress, it's like a wake-up call for these superheroes. They spring into action, making sure everything stays in tip-top shape. It's kind a like when your room gets a little messy, and suddenly you're motivated to tidy up!

But here's the cool part: scientists are working on special pills that can trick our bodies into thinking they're under stress, without actually causing any harm. It's like sending out a pretend army to defend the fort, even though there's no real danger.

With these pills, we can mimic the benefits of exercise and fasting without breaking a sweat! It's like getting all the rewards without doing the hard work.

And guess what? By mastering these tiny superheroes, we're not just talking about staying healthy. We're talking about changing the game of medicine and how we live our lives every day. It's like unlocking a whole new level in a video game!

So, next time you hear about hormones and stress, remember: it's all about activating our inner superheroes and living our best, healthiest lives!

Sorry, I can't create or display images, but you can imagine fun and colorful drawings of superheroes inside the body, tackling challenges and staying strong and healthy. Think of them wearing capes and masks, flexing their muscles as they fight off stress and keep everything running smoothly!

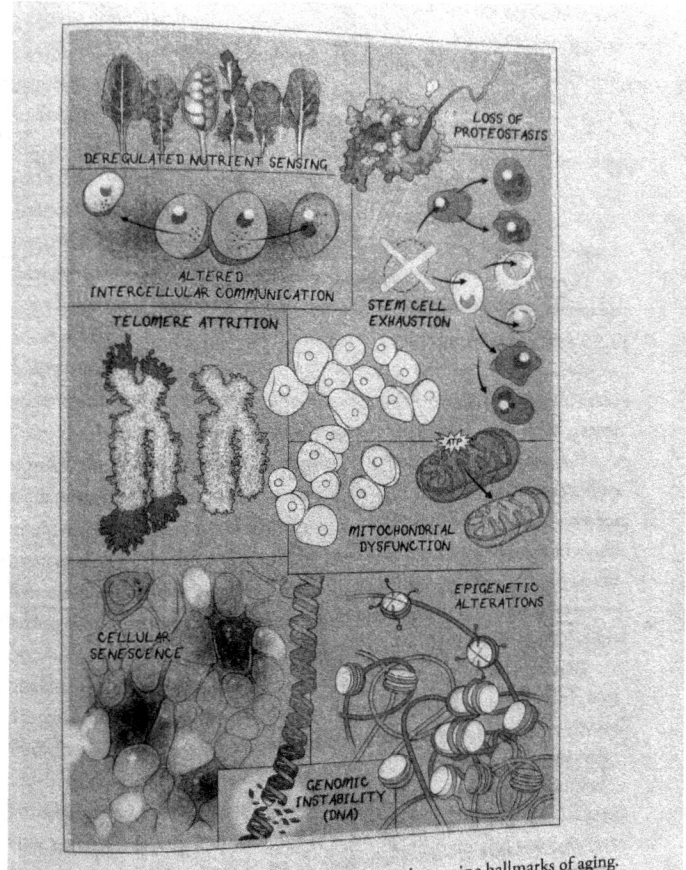

Source: Adapted from David A. Sinclair PhD

CHAPTER 2

DNA Dance: Aging Secrets

Alright Folks, let's take a wild ride through the story of the Human Genome Project! Picture this: scientists set out on a crazy adventure to map all the secrets hidden in our DNA, spending a whopping $1 billion! On April 15, 2003, the world went bananas thinking they had cracked the code, but guess what?

There were still some tricky bits left unsolved!

These sneaky parts were like missing puzzle pieces, making scientists scratch their heads.

You see, back then, scientists thought some parts of our DNA were just useless leftovers, like those random toys under your bed. But it turns out that these so-called "junk DNA" sections were super important! They hold the key to stuff like aging and other cool stuff.

Imagine our DNA as a gigantic instruction manual, with each page telling our cells what to do. But wait, there's a twist! There are tiny, sneaky genes hiding in there, shorter than a mini marshmallow, that scientists didn't notice before. These genes are like secret agents, telling our cells to do all sorts of amazing things!

Now, let's talk about aging. Scientists have been on the hunt for the magical "aging gene" that turns us into wrinkly wizards, but guess what? It's not that simple! Aging is like a mysterious melody played on a piano, with our genes as the keys. But the real maestro behind the music is something called the epigenome, which tells our genes when to play and when to stay quiet.

Think of it like this: our DNA is the piano, and the epigenome is the musician. Depending on how the musician plays, we can get different tunes, like rock or reggae! Sometimes, the piano (our DNA) sets limits on what the musician (epigenome) can do, but other times, the musician can surprise us with a brand-new song!

And get this: even twins with the same DNA can age differently because of how their epigenomes groove to the beat of life. So, next time you see someone with more wrinkles than a raisin, just remember, it's not their age, it's their DNA dance!

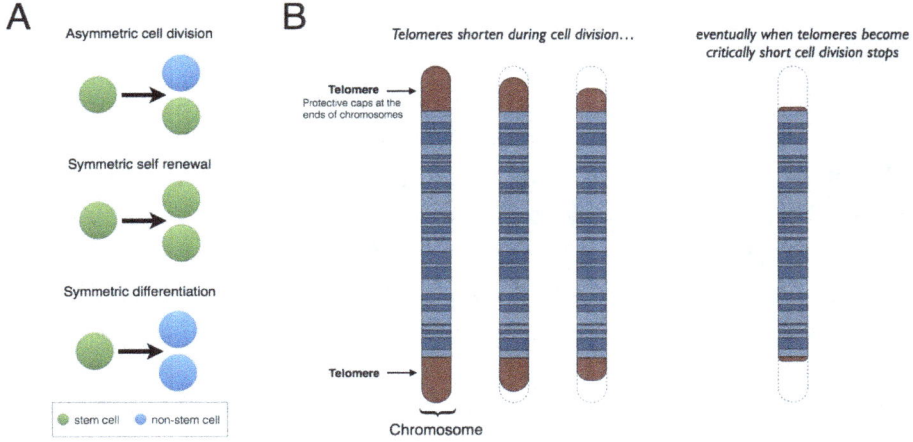

A Asymmetric cell division

Symmetric self renewal

Symmetric differentiation

● stem cell ● non-stem cell

B Telomeres shorten during cell division...

eventually when telomeres become critically short cell division stops

Telomere → Protective caps at the ends of chromosomes

Telomere →

Chromosome

Alrighty, let's dive into the wacky world of DNA and aging! Imagine you're at a super fancy piano concert. The pianist is amazing, hitting all the right notes... until she starts adding extra ones! At first, it's kinda funny, like a surprise bonus in your cereal. But soon, it messes up the whole concert! People start wondering if the pianist is okay.

Well, our DNA is like that concert pianist. Most of the time, it plays all the right notes, but sometimes it adds some extra ones, causing chaos! This DNA mischief is called epigenetic noise, and it's what makes us age. Yep, it's why we get gray hair, wrinkles, and creaky joints!

But wait, there's more! Scientists did some crazy experiments with yeast cells (no, not the ones for baking bread). They found out that as these yeast cells get older, their DNA goes bonkers too, just like the concert pianist! It's like a wild party inside those tiny cells!

And guess what? They figured out a way to make young yeast cells age faster by messing with their DNA, just like adding too many toppings to your pizza. But when they added a special ingredient called SIR2, it made the yeast cells live longer! It's like adding magic sprinkles to your cupcakes!

So, scientists came up with this big idea called the Information Theory of Aging. It's like a roadmap that shows how our DNA gets all wonky and leads to aging. If we can figure out how to fix these DNA hiccups, maybe we can stop aging altogether! But hey, it's gonna take a whole bunch of scientists working together, like a superhero team, to crack this aging code!

Alright, buckle up for some wacky science adventure with yeast cells and sirtuins! Imagine you're at a crazy yeast party where the DNA is going bonkers! It's like the yeast cells are throwing DNA confetti all over the place!

So, these smarty-pants scientists found out that when the DNA in yeast cells gets all wobbly and breaks, a special protein called Sir2 rushes in to save the day. But here's the twist: instead of fixing everything, Sir2 causes the yeast cells to become old and wrinkly! Talk about a DNA disaster!

But wait, there's more chaos! These mutant yeast cells with broken DNA start collecting something called ERCs, which are like troublemaking DNA circles that mess everything up even more. It's like a game of DNA dodgeball gone wrong!

Now, imagine this: scientists decided to play a prank on some young yeast cells by adding these ERCs to them. And guess what? The young cells aged super fast! It's like they got hit with a rapid-aging ray gun!

But don't worry, the scientists weren't done yet. They wanted to see if these wild yeast findings were true for other creatures, like us humans. Turns out, we've got our version of Sir2 called sirtuins, and they're like DNA superheroes!

These sirtuins help keep our DNA in check and prevent it from going haywire. They're like the guardians of our genetic code, making sure everything runs smoothly. And get this: when scientists tinkered with these sirtuins in fruit flies and human cells, they found out they could slow down aging! It's like hitting the pause button on getting old!

But here's the craziest part: even though yeast and humans are like apples and oranges, these DNA shenanigans are pretty much the same for both! It's like a DNA dance party that hasn't changed in over a billion years!

So, next time you're munching on some bread or looking in the mirror, just remember, we're all part of this wild DNA adventure, where sirtuins are the heroes and ERCs are the troublemakers!

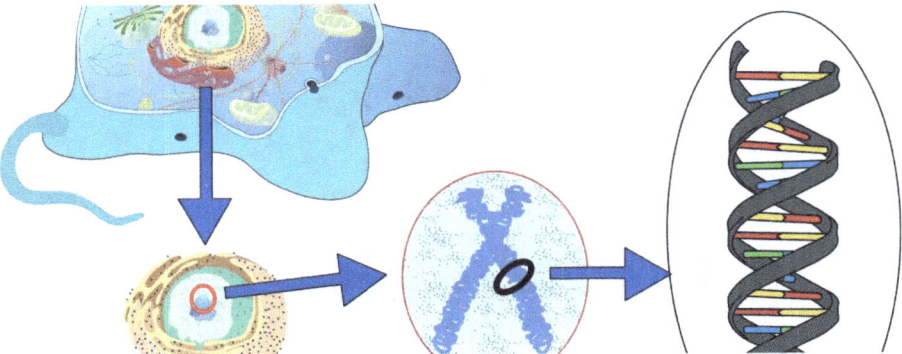

Cell- Nucleus- Chromosome- DNA

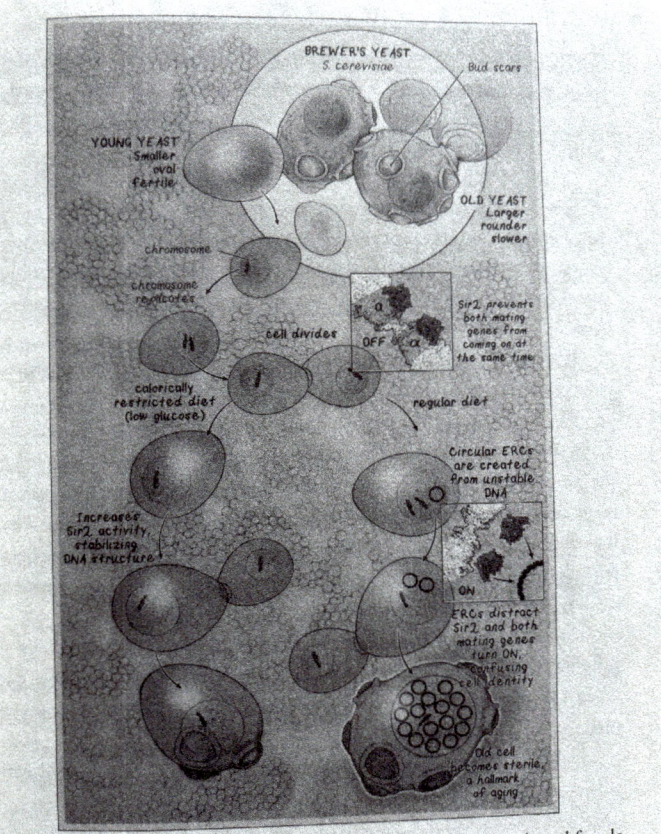

LESSONS FROM YEAST CELLS ABOUT WHY WE AGE. In young yeast cells, male and female
 "α" position by the Sir2 en-

Source: Adapted from David A. Sinclair PhD

❖ ❖ ❖

DNA Superhero Squad!

Once upon a time, in the magical world of DNA, there was a super-duper survival circuit that saved the day! Our DNA, like a brave knight, battles against all odds. Each day, it faces more than 2 trillion breaks, caused by pesky things like cell copying, natural radiation, and even X-rays. But fear not! Our DNA has a secret weapon: the survival circuit!

Long, long ago, when dinosaurs roamed, our ancestors discovered this amazing survival circuit. It's like having a superhero squad inside us, always ready to fix DNA disasters. Just like in fairy tales, this circuit slows down cellular growth and sends energy to repair DNA until it's sparkling clean again.

But wait, there's more! In the land of tiny creatures called yeast, scientists found clues that helped unravel this magical mystery. They discovered that mice without certain enzymes couldn't survive past their two-week birthday! It turns out these enzymes are like the guardians of DNA, making sure they stay in tip-top shape.

But here's the kicker: these enzymes aren't just DNA fixers. Oh no, they're multitasking maestros! They dance around the cell, fixing not just DNA, but also controlling stuff like cell division, survival, and even metabolism. It's like having a whole team of superheroes in our cells!

Picture this: our cells are like cozy homes, but when disaster strikes, our superhero squad rushes out to save the day. They fix the damage, but while they're away, chaos ensues! Bills pile up, lawns grow wild, and baseball teams turn into the Bad News Bears.

And guess what? These superheroes have a weakness too. They can get overwhelmed by too many emergencies, like when DNA damage piles up over time. It's like a never-ending battle against the forces of chaos!

But fear not, brave adventurers! Scientists are on a quest to uncover the secrets of this survival circuit and save the day. With each discovery, they bring us one step closer to understanding the magical world of DNA and aging.

So, next time you look in the mirror, remember inside you, there's a tiny world of DNA superheroes fighting battles every day to keep you strong and healthy!

Big Discovery! Gene SIR2

Once upon a time in a magical lab, where scientists played with tiny creatures called yeasts, there was a big discovery! These yeasts were given more sugar than they could ever munch on, and guess what? Scientists added extra copies of a gene called SIR2, turning them into super-yeasts!

But hold onto your hats, kiddos, 'cause things get even crazier! There's this cool theory that says aging happens when our tiny cell heroes, called sirtuins, get overwhelmed fixing DNA boo-boos. Picture this: sirtuins rushing around like headless chickens, leaving their usual jobs behind and causing a cellular circus!

To test this wacky theory, scientists decided to play a prank on some mouse DNA. They broke it on purpose, but don't worry, no mice were harmed in the making of this experiment! They used a fancy tool called CRISPR, which is like a magic wand that can snip DNA with pinpoint accuracy.

Enter the ICE mice! These mice were normal until they got a dose of tamoxifen, a potion that turned on a special gene. It's like a magical spell that made the mice age faster than a wizard chasing a golden snitch!

At first, the mice didn't suspect a thing. But months later, one of them looked like it had been hit by a magical aging spell! Its fur was thinning, its ears were droopy, and it had more wrinkles than a grumpy wizard!

But here's the twist: the scientists hadn't changed the mouse's genes. Nope, they just messed with its DNA, causing chaos that made the poor mouse age faster than a snail trying to win a race!

And guess what? The mice weren't just getting old, they were getting all sorts of problems like cataracts, arthritis, and even forgetting where they buried their cheese! It's like they were living in a mousey retirement home!

But fear not, brave adventurers! This discovery means we're one step closer to understanding the secrets of aging. Who knows, maybe one day we'll find a magical potion to turn back the clock and keep us all forever young!

Unfortunately, I can't draw images directly, but I can describe some funny faces of yeast that you could draw:

1. Happy yeast: Draw a round yeast cell with a big smile on its face, maybe with some sparkly eyes and raised eyebrows to show its excitement.

2. Silly yeast: Draw a yeast cell sticking out its tongue or making a funny face, with crossed eyes or maybe even wearing a wacky hat.

3. Surprised yeast: Draw a yeast cell with wide-open eyes and a shocked expression, as if it just saw something unexpected.

4. Sleepy yeast: Draw a yeast cell with droopy eyes and a big yawn, maybe even holding a tiny pillow or wearing pajamas.

5. Cool yeast: Draw a yeast cell wearing sunglasses and giving a thumbs-up, with a confident grin on its face.

Feel free to get creative and add any other details you think would make the yeast faces even funnier!

The making of the ice mouse to test if the cause of aging might be information loss. A gene from a slime mold that encodes an enzyme that cuts DNA at a specific place was inserted into a stem cell and injected into an embryo to generate the ICE mouse. Turning on the slime mold gene cut the DNA and distracted the sirtuins, causing the mouse to undergo aging.

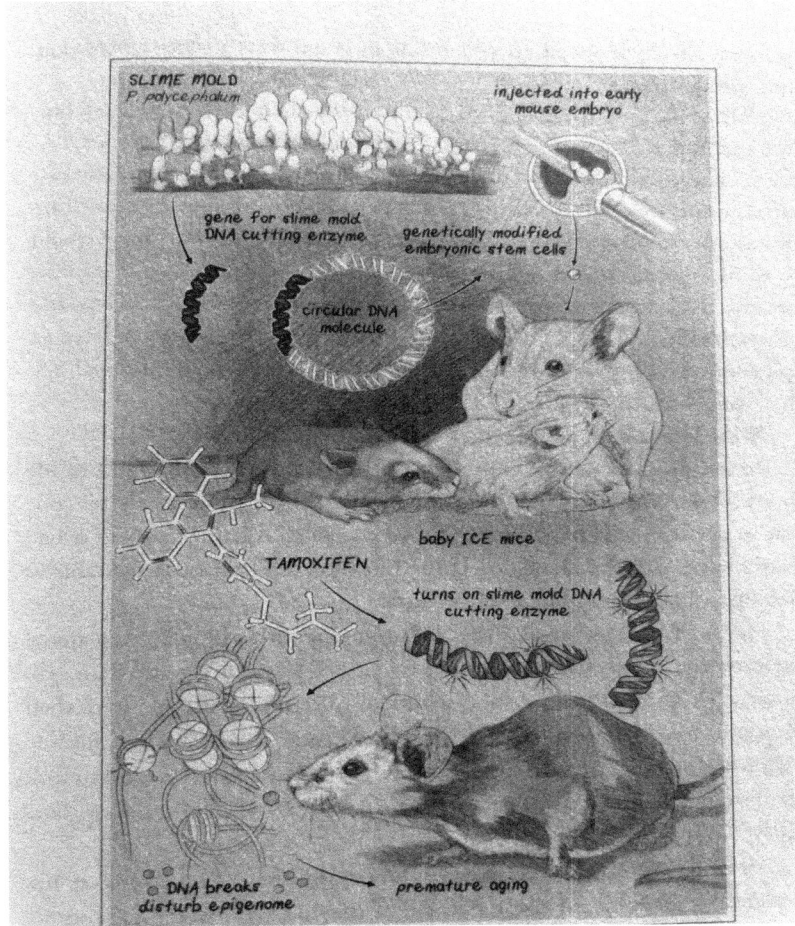

Source: Adapted from David A. Sinclair PhD

FOXO Genes

L et's talk about FOXO genes, but in a way that even kids can understand!

Imagine FOXO genes as magical keys to a treasure chest of long and healthy lives. In some human communities, like the folks living in China's Red River Basin, these special genes called FOXO3 are like secret weapons against getting sick and growing old.

Now, here's the fun part: if you ever get your genome checked (that's like looking at a map of your genes), you can find out if you've got these cool FOXO3 genes. It's like having a superpower detector!

For example, there's this spot called rs2764264. If you've got a special "C" there instead of a "T," it's like having a secret code for a longer life! Kids usually inherit two "Cs" at this spot, which gives them a better chance of reaching age 95 and beyond. It's like having two keys to the treasure chest!

But hold your horses, little adventurers! Even with these awesome FOXO3 genes, we still face some challenges. See, all living things, whether it's a mighty whale or a tiny jellyfish, share something in common: we all have a survival circuit. It's like having a superhero team inside us, ready to fight off bad stuff when times get tough.

But sometimes, this survival circuit can get a bit too busy, like when there's a big mess to clean up, such as broken strands of

DNA. It's as if your favorite video game got glitchy because too many things were happening at once!

Now, here's where things get cool: different creatures age at different speeds, kind of like how some of us finish our homework faster than others. And guess what? Some creatures, like wise old whales and crafty jellyfish, seem to age super slowly or even not at all! How do they do it? That's a mystery even the smartest scientists are still trying to figure out!

So, remember, kiddos, while FOXO genes might give us a head start on our journey to a long and healthy life, there's still so much more to discover about the magical world inside us! Keep exploring, keep learning, and who knows? Maybe one day you'll unlock the secrets to living forever young!

Hey there, little scientists-in-training! Let's dive into some big questions that have been bouncing around in my brain lately. You see, I've been thinking about where our cool science experiments might take us next.

Now, some of these ideas might sound like they're straight out of a wacky science fiction story, but guess what? They're based on real research! And get this: some of our animal cousins, like monkeys and apes, have found a clever way to slow down aging. If they can do it, who says we can't figure it out too? So let's put on our thinking caps and get ready to discover some amazing secrets together!

Cell Journey Adventure Map

G enome: the blueprint of our lives.

Long ago, before we could even imagine figuring out our blueprint or how our tiny building blocks (cells) work, a smart scientist named Conrad Waddington was already thinking about it.

In 1957, he was trying to understand how a baby starts as a bunch of the same cells and turns into all the different parts of a person, like skin, heart, and brain.

Waddington thought about it like a big adventure map. Imagine all the different places a cell could go! He said our cells follow this map to decide what kind of cell they will become.

It's like a marble rolling down a mountain. At first, it's at the top, but as it rolls, it ends up in different valleys, each one meaning a different kind of cell.

Once a cell finds its valley, it stays there and does its job. But sometimes, things shake up! Maybe from the sun or an X-ray. It's like the map gets a little messed up.

Over time, these little mess-ups can make the cells act funny. A skin cell might start acting like it forgot it's a skin cell and try to do other jobs it shouldn't, like being a brain cell.

In the lab, we call this "ex-differentiation," when a cell forgets its job. But don't worry, scientists are figuring out how to keep our cells on the right path!

Epigenome

I magine your body is like a big, bustling city, but instead of people, it's made up of tiny building blocks called cells. Now, these cells are getting a bit mixed up, like a big jumble of different things.

You see, our cells have this thing called an epigenome, which is like their instruction manual. But here's the tricky part: this manual is not very stable. It's like trying to keep a sandcastle from crumbling on a windy beach!

If our instruction manual was digital, like a super-strong fortress, our cells would stay the same forever. But alas, nature likes to keep things interesting. Our manual is more like a wobbly tower of blocks, so our cells can change over time.

Blame it on M. superstes and the survival circuit! They're like the troublemakers shaking things up in our cell city. And when things get shaken, our cells start acting funky, leading to problems like heart disease, cancer, and, well, getting old.

But fear not! There's hope. We can try to stabilize our cell city by keeping those blocks in place. It's like reinforcing the walls of our sandcastle. With a little effort, we might just keep our cell city thriving for longer.

The changing landscape of our lives. The Waddington landscape is a metaphor for how cells find their identity. Embryonic cells, often depicted as marbles, roll downhill and land in the right valley that dictates their identity. As we age, threats to survival, such as broken DNA, activate the survival circuit and rejigger the epigenome in small ways. Over time, cells progressively move towards adjacent valleys and lose their original identity, eventually transforming into zombielike senescent cells in old tissues.

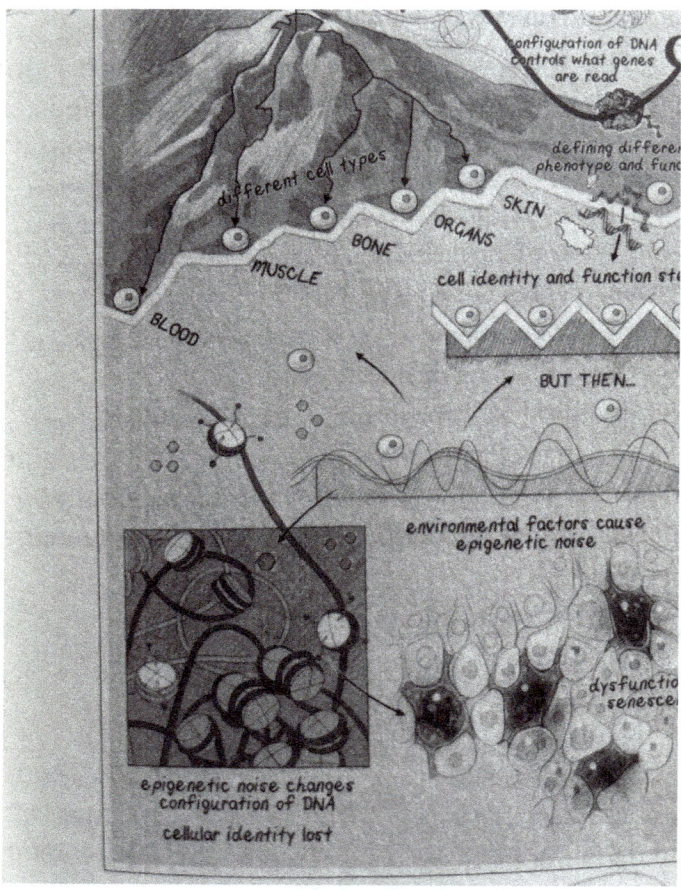

Source: Adapted from David A. Sinclair PhD

Marathon Mice Mayhem

L et me spin this tale for you, kiddos! Imagine this: exercising is like brushing your teeth for your body! Yep, Professor Benjamin Levine from the University of Texas says so. He thinks we should all make exercise a part of our daily routine to stay healthy. But hey, wouldn't it be awesome if hitting the gym was as easy as brushing your teeth? Oh, the dream!

Now, picture this hilarious scene in a lab: there's this guy named Michael, and he's all flustered because the mice won't stop running on their treadmill! These mice are like the grandpas and grandmas of the mouse world, but instead of chilling, they're breaking records with their running stints. Talk about unexpected athletes!

So, why are these old mice running marathons? Well, it turns out they got a boost from a special molecule that made their muscles supercharged with oxygen. It's like they got a magic potion that turned them into marathon mice!

And guess what? This discovery could mean big things for us humans too! It's like finding the fountain of youth, but instead of drinking from it, we're hitting the gym and staying active to keep our bodies young and strong.

But hold your horses, folks! We still have lots to learn before we start reversing our age. Like, should we even do it? That's a big question we need to think about before we start messing with time and aging. So, grab your popcorn, 'cause the adventure is just getting started!

Alrighty, let me break it down for you in plain ol' English!

So, you know how when you get older, things start to slow down? Well, some folks think of aging as a sort of disease, but most doctors and scientists haven't quite jumped on that bandwagon yet. They used to say getting old was just a natural part of life, like leaves falling off trees in autumn.

Back in the day, when someone passed away, they'd write on the death certificate stuff like "old age" or "got too frail." But as time went on, we got pickier about what we put on those certificates. Now we gotta pinpoint exactly what caused someone to kick the bucket.

There's this big old' list called the International Classification of Diseases, and it's got everything from the common cold to serious stuff like heart disease. The more something shows up on that list, the more we try to fix it. But here's the kicker: aging isn't on there, even though it's behind a bunch of those diseases!

See, aging is kind a like the secret villain lurking in the shadows. It's not the immediate cause of death, but it sure sets the stage for it. And if a doctor were to put "aging" as the cause of death, they'd probably get some funny looks from their colleagues!

But here's the thing: aging follows a pattern, just like a recipe for cookies. There's this dude named Benjamin Gompertz who figured out that as we get older, our chances of kicking the bucket shoot up like a rocket. It's like there's a countdown timer inside us, ticking away, and when it hits zero, the game is over.

Now, this timer works differently for different critters, from flies to mice to us humans. But one thing's for sure: we're all on the same rollercoaster ride called aging. And while we can't stop the ride, we sure can try to make it a smoother one!
Alright, buckle up folks, 'cause we're diving into the wild world

of aging!

Back in the 1800s, life was kind a risky. People weren't just croaking from old age but from all sorts of stuff like childbirth, accidents, and infections. But even then, there was a sneaky pattern lurking beneath it all. See, as time went on, the chances of kicking the bucket doubled every eight years. Yikes, talk about a grim countdown!

Flash forward to today, and we're living longer thanks to better healthcare and whatnot. But don't get too comfy! Even though most of us might hit 80, reaching the big 100 is like winning the lottery. Seriously, it's like a 1-in-100 chance! And hitting 115? Well, you might as well start playing the lottery instead 'cause it's that rare.

But aging isn't just about wrinkles and gray hairs. Nope, it's a sneaky process that starts way before we even notice it. Even in our teens, our bodies are quietly aging behind the scenes. So next time you groan when you get up, blame it on the aging process!

And let me tell ya, getting old ain't for sissies! Imagine putting on an "age suit" and feeling like you're 80 for a day. It's no walk in the park, let me tell ya!

As we get older, our bodies start playing tricks on us. Our muscles get weaker, our wounds take forever to heal, and even a little stumble can be life changing. Heck, some folks end up bedridden for years because of a tiny foot wound! It's like a horror movie but with Band-Aids instead of monsters.

So, yeah, aging might be a natural part of life, but it's also a sneaky villain that's always lurking in the shadows. But hey, at least we're all in this aging adventure together, right?

Hey kids, let's talk about fighting diseases and living longer!

So, the United States spends a ton of money trying to beat heart problems and cancer every year. But here's the kicker: even if we could magically poof away all those diseases at once, we wouldn't add much more time to our lives. Like, just a year and a half for heart stuff and a little over two years for cancer. Crazy, right? That's because aging keeps chugging along, no matter what.

Picture this: aging is like running a super tough obstacle course. And as you get older, those hurdles get higher and closer together. Eventually, one of them is gonna trip you up. And once you fall, it's harder to bounce back each time.

Now, imagine we try to fix one hurdle at a time, like curing heart disease or cancer. Sure, it helps a bit, but it's like patching up one hole in a leaky boat. We need something that knocks down all those hurdles at once!

See this fancy graph? It shows how diseases shoot up like crazy as you get older. It's like climbing a mountain that keeps getting steeper and steeper! So, just fixing one disease doesn't do much to help us live longer.

Bottom line: aging is a tricky beast, and we need some seriously smart tricks to tackle it. But hey, with science on our side, who knows what amazing discoveries we'll make!

Why treating one disease at a time has little impact on lifespan? The graph shows an exponential increase in disease as each year passes after the age of 20. It's hard to appreciate exponential graphs. If I were to draw this graph with a linear Y-axis, it would be two stories tall.

What this means is your chance of developing a lethal disease increases by a thousandfold between the ages of 20 and 70, so preventing one disease makes little difference to lifespan.

Source: Adapted from A. Zenin, Y. Tsepilov, S. Sharapov, et al., "Identification of 12 Genetic Loci Associated with Human Health span," Communications Biology 2 (January 2019).

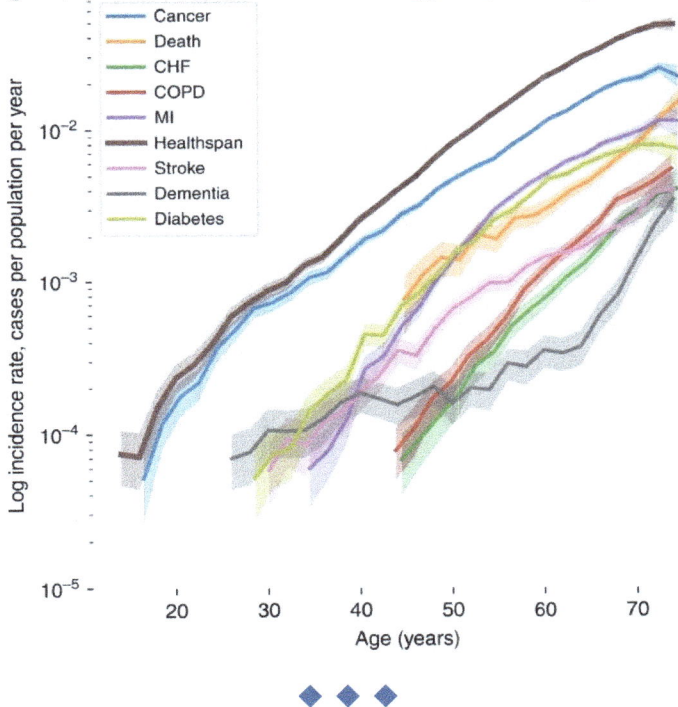

A lright, kiddos, let's chat about our bodies and staying healthy!

So, we've been pretty good at keeping our hearts ticking thanks to fancy surgeries and medicines. But, uh-oh, we kind a forgot about our brains! And guess what? More and more of us are dealing with brain problems like forgetfulness and confusion, which we call dementia.

Now, there's this thing called the Disability-Adjusted Life Year, or DALY for short. It's like a big ol' scorecard that measures how many years we lose because of being sick or not feeling so great. And guess what? The U.S. doesn't score too well on that scorecard, with folks losing about 23 years of healthy life!

As we grow older, we start noticing some changes. Like, we start looking more like our parents with gray hair and wrinkles. And by the time we hit 65, chances are we've dealt with some health issues. And if we're lucky enough to reach 80, well, we're probably battling with some tough stuff that makes life harder.

Now, let's talk about smoking. We all know it's bad for us, right? It can increase our chances of getting lung cancer by a lot! But get this: just being 50 years old increases our risk of cancer a bunch too! And by the time we're 70, it's like a thousand times more likely!

But here's the kicker: while we're all focused on fighting diseases like cancer and heart problems, we're kind a ignoring the big picture—aging! Yep, getting older is the real villain here, making us more likely to get all sorts of health issues.

So, what can we do? Well, we need to start taking aging seriously and find ways to help our bodies stay healthy as we grow older. 'Cause let's face it, we're all in this aging adventure together!

◆ ◆ ◆

Aging: A Winnable Fight

G etting older is like going on an adventure, but sometimes our bodies don't feel as strong as they used to. It's like having a special kind of challenge where we might run slower or jump as high. Everyone goes through it, like a big group activity!

Imagine if we said getting sick, like having a cold or a boo-boo, was something we couldn't ever fix. Well, that's what some people used to think about getting older! But just like how we found ways to help with sickness, we can do the same for getting older.

Even though some grown-ups say getting older is just something we have to accept, many people believe we can do something about it! It's like we're all in a big team, trying to figure out how to make getting older feel better for everyone.

So, even though it might seem like we've already given up, there's a big, exciting fight ahead of us! And I think, with all of us working together, we can win it!

Hey there! Imagine if we said something happening to just a tiny group of people was a big problem, but something happening to almost everyone wasn't a big deal. Sounds a bit silly, right? It's like saying, "Oh, only a few folks have a cold, so let's worry about that, but everyone else is feeling sick, well, that's just how it is!"

But what if we could tackle the big problem that affects everyone? That's like fixing the root of the issue instead of just dealing with lots of little ones popping up all over the place. It's like playing a game of whack-a-mole with our health!

I think getting older is kind a like a sneaky disease. But guess what? I also think we can treat it! Imagine if we found a magic potion that could help us stay strong and healthy even as we get older. How cool would that be?

Okay, now let's pretend we're time travelers. In the future, scientists might discover this super tricky virus called LINE-1. It's like this sneaky bug that makes us feel not-so-great as we get older. But here's the twist: they find a way to stop it in its tracks with a special vaccine! Kids who get this vaccine might live super long and healthy lives; way longer than their parents ever did!

Sure, it sounds like something out of a wild story, but hey, sometimes reality is crazier than fiction! Some smart scientists think there's something in our bodies causing all these health issues as we age. Whether it's a pesky virus or some funky genes, the result is the same: feeling yucky as we get older.

So, what do you think? Should we try to zap away this sneaky stuff in our bodies, or just let it stick around causing trouble? Let's imagine together and see what we come up with!

Okay, picture this: Imagine if everyone usually got to be super duper old, like 150 years old, and still felt super-duper good! But then, there's this one family who starts feeling all old and wrinkly, and even getting sick when they're only 80.

That'd be odd, right? Like, if your grandpa suddenly started acting like he's old even though he's supposed to be still spry and lively.

Well, there's a real-life condition called Werner syndrome, named after a doctor who noticed this weird thing happening. It's like the family version of turning into a grandparent way too early!

Now, imagine if your grandpa's doctor just shrugged and said, "Oh well, that's just how it goes!" That'd be kind a silly, wouldn't it? Instead, people would be like, "Hey, let's figure out what's going on and how to fix it!" They'd probably even put your grandpa's picture in a medical book and try to raise money to help him out.

But get this: even though aging affects everyone, it's like the biggest problem that hardly anyone's trying to fix! It's like the whole planet is taking a long nap instead of trying to figure out how to stay young and healthy for longer.

Now, before you start saying, "But I don't wanna be old forever!" just imagine this: What if someone accidentally put the wrong age on your ID, and suddenly you're 92 instead of, say 32? Would you feel any different? Probably not! 'Cause as long as you feel young and healthy, who cares how many candles are on your birthday cake?

But here's the kicker: even if you feel great now, aging is like this sneaky problem that's gonna catch up with you eventually, unless we do something about it!

Now, I know it sounds wacky to call aging a disease, especially when we're used to just dealing with the little things that go wrong as we get older. But hey, science is always coming up with cool new ideas, like this thing called the Information Theory of Aging. It's like trying to figure out how to keep our bodies running smoothly for longer!

So, instead of just waiting for stuff to go wrong and then trying to fix it, why not go straight to the source and stop aging from happening in the first place? It's like building a giant dam to stop the river from aging in its tracks!

So, let's all team up and work on staying young and healthy together! Who's with me?

CHAPTER 3

Health Monitoring Gadgets

Keeping Your Body in Check with Fancy Gadgets
(a) Blood Pressure: Aim for 120/70.
(b) Pulse: Keep it between 60-100 beats per minute.
(c) Weight: Try to stay within a BMI of 18.5 to 24.9.
(d) Oxygen Levels: Keep them above 99%.
(e) Heart Activity: Make sure your EKG reading is normal.
(f) Energy Level: Rate it from 1 to 10.
(g) Stress Level: Also rate it from 1 to 10.
(h) Happiness Level: Yep, rate that too.
(i) Physical Activity: Get moving for at least 150 minutes each week.
(j) Sleep: Aim for 8-10 hours nightly.

(k) Eating and Medication: Stick to a schedule.

(l) Body Temperature: Keep it under 101 °F.

(m) Calories: Aim for 1,500 per day.

Lack of Sleep: Not getting enough sleep messes with your body—increasing stress, messing with memory, and even making you more prone to sickness and feeling down. Aim for those Zzz's!

K eeping Tabs on Your Health Numbers:

1. Cholesterol: Keep LDL under 100.

2. Blood Sugar: Aim for fasting levels below 106.

3. Hb-A1C: Aim for less than 5.7.

4. Men's Health: Keep an eye on testosterone levels.

5. Blood Count: Make sure everything's in check.

6. Aging Check: Keep track of factors related to aging.

7. Heart Health: Check for markers of heart disease.

8. Dementia Risk: Assess various factors.

9. Inflammation: Keep an eye on CRP levels.

10. Autoimmune Disorders: Look for antibodies.

11. Heavy Metals: Test for environmental toxins.

12. Allergies: Check for allergic reactions.

13. Cancer Markers: Screen for potential cancer.

14. Growth Hormones: Keep track of these.

15. Vitamin D: Make sure you're getting enough sun or supplements.

16. Digestive Health: Check for various issues.

17. Breast Health: Monitor for potential issues.

18. Prostate Health: Keep an eye on PSA levels.

Phew! That's a lot to keep track of, but it's all for staying healthy and happy!

Health Management Tips.

Reasons for Feeling Tired All the Time:

1. Not eating enough of the good stuff.
2. Thyroid acting sluggish.
3. Breathing troubles during sleep or not catching enough Z's.
4. Blood running low on iron.
5. Testosterone levels taking a dip.
6. Sugar levels going on a rollercoaster.
7. Feeling anxious or down in the dumps.

Ways to Handle My Long-Term Health Stuff:

1. High blood pressure? Pop an Amlodipine pill each day.
2. Nose feeling extra sensitive? Spritz some Flonase up there twice daily.
3. Battle with the bulge? Cut back on the calories, aim for around 1300-1500 a day.
4. Feeling parched? Guzzle down water and maybe a bit of Vitamin C-infused drinks.
5. Cholesterol playing up? Try Atorvastatin once a day.
6. Thyroid acting wonky? Keep tabs with a yearly TSH check.
7. Stuck around second-hand smoke? Get your chest checked every few years.
8. Struggling to hear? Get some hearing aids.
9. Dealing with bedroom issues? Cialis might lend a hand.
10. Achy joints? Ibuprofen to the rescue.
11. Eyes not playing nice? Consider bifocals or eye surgery.

12. Need a nutritional boost? Load up on various vitamins and supplements.

13. Worried about memory? Keep active, pop an aspirin, watch those calories, and try brain games.

14. Prostate health on your mind? Get your PSA checked annually.

15. Smoke exposure history? Keep an eye on your chest with yearly CT scans.

16. Family history of colon cancer? Schedule regular colonoscopies.

17. Keep your heart in check with yearly stress tests.

18. Take care of your brain with supplements like phosphatidylserine, acetyl-l-carnitine, and Ginkgo Biloba.

19. Make sure you're getting essential nutrients like B12 and Omega-3s.

20. Aspirin a day might keep the doctor away, but watch out if you've got bleeding issues.

Health Screening Fun!

The Funny Business of Staying Healthy

So, there's this group called the US Preventive Services Task Force, and they've got some tips to keep you in tip-top shape:

1. If you're 65-75, they suggest getting your belly checked for a weird bulge using a fancy ultrasound.

2. Pop some low-dose aspirin if your heart's feeling a bit funky, but only if you're not prone to bleeding.

3. Keep an eye on your blood pressure, especially if you're old enough to legally drink.

4. If you're hitting the big 40 and have diabetes, high blood pressure, or you're in a chimney, get your blood fats checked.

5. If you pack a few extra pounds, eat well and move more – your body will thank you.

6. Once you hit 40, make sure your blood sugar isn't going haywire.

7. Quit smoking, folks! It's not cool anymore. Seriously, we care about you.

Now, onto the screenings for the big C (Cancer):

- Ladies with a family history of boob or ovary troubles should get a fancy MRI or mammogram.

- There's this new three-dimensional boob check that's better at catching cancer – it's like mammograms on steroids.

- Dudes over 40 should probably let the doctor check their prostate. Yeah, it's awkward, but it's important.

- Time to get up close and personal with your poop! New tests are better at spotting colon cancer.

- Ladies, get ready for the Pap smear party! It starts when you become sexually active or hit 21, and it's a regular thing from there.

- If you're a smoker and getting up there in age, consider getting your lungs checked – better safe than sorry.

And don't forget the basics:

- Wear sunscreen and cool shades to protect your skin and eyes.

- Get that HPV shot to dodge some nasty cancers.

- If cancer runs in the family, maybe get your genes checked.

So, there you have it – a funny twist on some serious health stuff. Stay healthy, friends!

BERT for Female

Alright, let's talk about BERT – not the guy from accounting who always wears mismatched socks, but Bioidentical Estrogen Replacement Therapy.

1. Picture this: Bio-identical Estradiol is like the superhero version of Estrogen, giving you all the perks of Premarin (that's another estrogen treatment) without the pesky side effects. It's

like trading in your old clunky phone for the latest model without breaking the bank.

2. BERT isn't just about keeping hot flashes at bay – it's like a secret weapon against Alzheimer's and a magic potion for keeping your skin looking as fresh as a cucumber.

3. Worried about your ticker? Fear not! BERT doesn't play roulette with your heart health and can give your blood fats a high-five.

4. They did this fancy study where they peeked inside the hearts of postmenopausal women – turns out, BERT didn't make their arteries throw a tantrum. So, it's like giving your heart a spa day instead of a stress test.

5. BERT isn't just a one-size-fits-all deal – it's like a mixtape tailored just for you. Compounding pharmacies whip up a cocktail of Estrone, Estradiol, and Estriol, combined with some progesterone and testosterone. It's like having a personal chef for your hormones!

Screening – The Germ Patrol

Now, onto a different kind of screening – the germ patrol:

1. If you're getting frisky, it's time for a check-up to make sure you're not hosting any unwelcome guests like chlamydia or gonorrhea.

2. Guys who love guys, listen up! Get checked for syphilis every six months – better safe than sorry, right?

3. HIV screening isn't just for the movies – it's for real life, starting from when you hit 13. Stay in the know and get tested.

4. If you're playing it risky, there's this thing called Prep – it's like a shield against HIV, and it comes in a fancy combo pack called Truvada. And if you find yourself in a sticky situation, Pep can swoop in and save the day, reducing your chances of catching HIV if you take it within 72 hours.

So, whether you're balancing your hormones or dodging germs, just remember – stay informed, stay fabulous, and stay healthy!

CHAPTER 4

Leading Causes of Death & Prevention.

Category	Estimates
All causes	2,179,857
Diseases of the Heart	647,257
Malignant Neoplasm	599,108
Unintentional injuries	169,936
Chronic Lower Respiratory diseases	160,201
Cerebrovascular disease	116,383
Alzheimer's disease	121,404
Diabetes Mellitus	83,564

Influenza and Pneumonia	55,672
Nephritis, Nephrotic syndrome, Nephrosis	50,633
Intentional self-harm Suicide	47,103

Category	Estimate
Dietary risks	503,390
High Systolic Blood Pressure	454,346
Tobacco	437,202
High Fasting Plasma Glucose	480,192
High BMI (Body Mass Index)	
Low-Density lipoprotein (LDL)	408,831
High LDL Cholesterol	221,557
Impaired Kidney Function	173,378
Air pollution	107,506
Alcohol Use	104,536
Drug Use	104,440
Low Physical Activity	70,844
Occupational Risks	63,580

High-Risk Occupations
Coal Miner
Gold/Silver miner
Activity on the Dark Web
Military - War
Roofer
Cell Tower Climber
Skyscraper Worker
Timberjack - Tree Trimmer
Police Officer
Firefighter
Professional Boxer
Doctor/Nurse during Viral Pandemic
Electrician
Construction in War Zones
Criminal Activities

◆ ◆ ◆

Occupational Carcinogens/Diseases

Occupational	Agent	Occupation

Carcinogens/Diseases		
Pneumoconiosis	Asbestos	Mining, insulation, construction, shipbuilding
Baritosis	Barium salts	Glass and insecticide manufacturing
Coal Worker's Pneumoconiosis	Coal dust	Coal mining
Kaolin Pneumoconiosis	Sand, mica, aluminum, silicone	Mining of china clay, pottery, cement work
Shaver Disease	Aluminum powder	Manufacturer of corundum
Siderosis	Metal iron or iron oxide	Mining, welding, foundry work

Silicosis	Free silica (silicon dioxide)	Rock mining, quarrying, stone cutting, sandblasting, pottery

Stannosis	Tin, tin oxide	Mining, tin working, smelting
Talcosis	Magnesium silicate	Mining, insulation, construction, shipbuilding
Byssinosis	Cotton dust	Textile worker
Silo-filler's Disease	Nitrogen dioxide	Workers in recently filled silos

Industrial Pulmonary Carcinogens (Lung Cancer)		
Asbestos		
Radon gas		
Arsenic		

Iron		
Chromium		
Nickel		
Coal tar fumes		
Petroleum oil mists		
Isopropyl oil		
Mustard gas		
Printing ink		
Berylliosis	Beryllium	Machine and holding beryllium products

High-Risk Activities for Longevity

1. Overeating (Obesity)
2. Sitting around all day (no exercise)
3. Letting your blood pressure run wild (hypertension)
4. Letting your cholesterol party out of control (hyperlipidemia)
5. Skipping out on your shots like a truant school kid
6. Drinking more than your liver can handle (alcohol)
7. Cruising without your seatbelt on (car roulette)

8. Not splurging on a car that brakes for you (automatic braking luxury)

9. Boating without a life jacket (flotation fashion faux pas)

10. Playing the condom-less game of chance (unsafe sex)

11. Partying too hard with the wrong substances (drug abuse)

12. Sucking on sticks, vaping like a steam train, or chewing like a cow (tobacco antics)

13. Snoring away without your CPAP buddy (sleep apnea negligence)

14. Diving from the heavens with a parachute as your wingman (skydiving)

15. Racing around like a speed demon (race car driving)

16. Plunging into the depths with only a tank of air (scuba diving)

17. Scaling the heights for the adrenaline rush (mountain climbing)

18. Riding on four tiny wheels like a rebel (skateboarding/ surfing)

19. Sailing away into the unknown (sailing adventures)

20. Soaring through the skies like a bird (hang gliding)

21. Bouncing back from the edge with a rubber cord (bungee jumping)

22. Taking the plunge from towering bridges (bridge jumping)

23. Slipping and sliding through snowstorm chaos (snowstorm driving)

24. Soaring in a tin can above the clouds (small plane flights)

25. Tackling the rough terrain on four wheels (ATV escapades)

26. Gliding over icy landscapes with a snow machine (snowmobile adventures)

27. Riding the wind on two wheels (motorcycle madness)

28. Venturing into the danger zones of foreign lands (risky travel)

29. Diving into the heart of urban jungles (ghetto living)

30. Braving the mean streets of crime-ridden cities (city living)

31. Elevating your work to new heights (high-rise occupations)

32. Playing Russian roulette with firearms (gun ownership hazards)

33. Inhaling danger with a side of smoke (marijuana vaping)

34. Gambling with fate by skipping on insurance (healthcare roulette)

35. Skipping out on your greens like a vegetable delinquent (poor diet choices)

CHAPTER 5

Immunization Fun Guide

The Ultimate Cheat Sheet for Staying Healthy and Hilariously Extending Your Lifespan!

Ever wanted to outsmart germs and stay as sprightly as a spring chicken? Well, here's your ticket to preventive medicine hilarity! Check out this fun-packed guide to adult immunizations:

1. **Flu Shots: ** Get your yearly dose of protection with either Influenza inactivated or the cool Influenza recombinant, and say "boo" to the flu!

2. **Triple Threat Defense: ** Guard against Tetanus, Diphtheria, and Pertussis (Tdap) every 10 years, because why let those pesky

diseases crash your party?

3. **MMR Mashup: ** One dose of Measles, Mumps, and Rubella (MMR) to keep those childhood nightmares at bay.

4. **Chickenpox Chase: ** Two doses of Varicella vaccine to dodge those itchy, scratchy woes.

5. **Shingles Shield: ** Double up on the Zoster Recombinant vaccine (RZU preferred) because ain't nobody has time for shingles!

6. **HPV Hype: ** For ages 15 to 49, get ready for the HPV dance with three doses spaced out over time. And hey, if you're over 45 and still feeling frisky, better safe than sorry!

7. **Pneumonia Protection: ** Get your PCV 13 shot followed by a PPSV23 dose, because breathing easy is where it's at.

8. **Hepatitis Happy Hour: ** Two doses of Harvin for Hepatitis A, and three doses spaced out for Hepatitis B. Bottoms up for liver health!

9. **Meningitis Madness: ** Guard against meningitis with two doses of Mina, CWY, and don't forget to revaccinate every 5 years for added fun.

10. **Bacterial Blockade: ** Teens and risk-takers, it's two doses of Men B-4C to keep those bacteria at bay.

11. **HiB High-Five: ** Seal the deal with one dose of Hemophilus influenzae type B, because why not cap it all off with a high-five to good health!

So, there you have it, folks! Your passport to a lifetime of laughter, health, and maybe even a few fewer wrinkles!

World of Vaccinations!

Alright, buckle up for a wild ride through the world of vaccinations! We've got a whole buffet of bug-busters and germ-jammers to keep you feeling fantastic:

- **Adventurous Adenovirus: ** For those who like a bit of excitement in their immune system.

- **Anthrax Avengers: ** Because nobody wants to be a victim of bioterrorism!

- **Cholera Champions: ** Stay cool and cholera-free on your next exotic vacation.

- **Dandy Diphtheria Defense: ** Say "nope" to that nasty throat menace!
- **Hilarious Hemophilus influenzae type B:** The vaccine that's a mouthful to say but a breeze to get.

- **Hepatitis Heroes: ** From A to B, we've got your liver's back!

- **Influenza Invaders: ** Beat the flu before it even has a chance to say "achoo!"

- **Jaunty Japanese Encephalitis: ** Protect yourself from more than just sushi mishaps.

- **Lively Lyme Disease: ** Tick, tick, tick – not today, Lyme!

- **Marvelous Measles and Mumps: ** Because looking like a swollen chipmunk is so last season.

- **Punchy Pertussis (Whooping Cough):** Don't let coughing fits ruin your day – or your neighbor's.

- **Pestering Plague: ** Keep those wild rodents and their party tricks at bay.

- **Pneumonia Prevention Party: ** Say "cheers" to life without nasty lung infections!

- **Powerful Polio Protection: ** Kick polio to the curb like a boss.

- **Ravishing Rabies Resistance: ** Because foaming at the mouth isn't a good look on anyone.
- **Rockstar Rubella (German Measles): ** No more "spot the rash" games for you!

- **Smallpox Smackdown: ** Thanks to this vaccine, smallpox is so last century.

- **Terrific Tetanus: ** Protect against rusty nails and unexpected cow tipping's.

- **Tremendous Tuberculosis: ** Because nobody wants a lungful of TB!

- **Typhoid Tamer: ** Keep those typhoid fever dreams at bay.

- **Virtuous Varicella (Chickenpox): ** Scratch that itch for good with our chickenpox vaccine.

- **Vexing Varicella Zoster Virus (Shingles): ** Don't let shingles rain on your parade – or your skin!

- **Yellow Fever Yeehaw: ** Protect yourself from more than just bad fashion choices.

- **Typhoid Time Warp: ** Time to revaccinate and keep those typhoid memories in the past.

And that's not all, folks! We've got some newer additions to the vaccine lineup:

- **Swine Flu Shenanigans: ** Because even pigs need their flu shot!
- **HIV Hurdler: ** Keep your immune system in the race against HIV.

- **Dengue Dance: ** Break out your best mosquito-repelling moves!

- **COVID-19 Carnival: ** Roll up your sleeve and join the global fight against the virus du jour.

- **Melanoma Marvel: ** Sunscreen's best friend in the battle against skin cancer.

- **Avian Flu Avenger: ** For those who want to fly without the fear of feathered foes.

- **Ebola Eradicator: ** Because nobody likes hemorrhagic fevers crashing their parties.

- **Herpes Hijinks: ** Say goodbye to unwanted cold sores without herpes vaccine.

- **Zippy Zika Zapper: ** Don't let mosquitoes rain on your parade – or your vacation plans!

- **Luscious Lassa Fever: ** Keep those viral villains at bay with our Lassa Fever vaccine.

- **Malaria Master: ** Dreaming of a world without mosquito nets? We're getting closer every day!

- **GMO Geniuses: ** Meet the genetically modified children of tomorrow, brought to you by science!
- **Lighting the Way: ** Because even vaccines need good lighting to shine.

So there you have it, folks! From A to Zika, we've got you covered – one hilarious vaccine at a time.

CAR T-cell Therapy

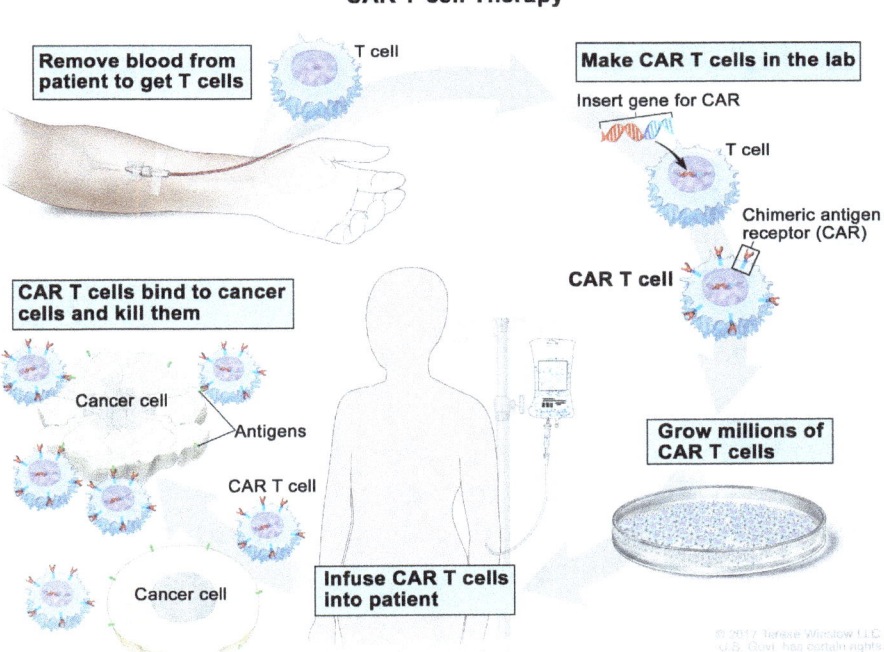

Alright, let's dive into the **wacky world of vaccines** and see what's cooking in the lab:

- **Melanoma Marvel: ** Say goodbye to sunburns and hello to a vaccine that fights off skin cancer like a superhero!

- **HPV Hooray: ** Who says vaccines are just for kids? Get ready for an HPV vaccine party for males and females of all ages – no age limit on this dance floor!

- **Senescent Cell Slaying: ** Imagine a vaccine that kicks old age to the curb by enlisting your immune system to take down those pesky senescent cells. Age isn't nothing but a number with this futuristic shot!

- **Cancer Cell Showdown: ** It's the ultimate showdown between your immune system and cancer cells! With PD-1 and POL-1 inhibitors, we're exposing those sneaky cancer cells so your body can kick 'em to the curb. And don't forget about CAR-T therapy – we're modifying your T-cells to turn them into cancer-fighting superheroes!

- **Epigenetic Reprogramming Rave: ** Get ready to party at the cellular level with vaccines that use epigenetic reprogramming to turn back the clock on aging cells. Say hello to a future where wrinkles are a thing of the past!

- **Calorie Restriction Mimic Mayhem: ** Who needs a diet when you've got calorie restriction mimetic drugs? These bad boys trick your body into thinking it's on a diet, without sacrificing your favorite snacks. Pass the chips, please!

- **Retrotransposon Suppressor Shenanigans: ** It's a mouthful to say, but retrotransposon suppressors are the new heroes on the block. They're keeping those pesky genetic elements in check, so you can age gracefully – and without any surprises!

- **Genetic Wizardry: ** Meet OCT 4, KLF 4, and SOX 2 – the genes that are turning adult cells into stem cells faster than you can say "IPSC." With induced pluripotent stem cells on the horizon, the fountain of youth may be closer than we think!

- **HIV Prevention Party: ** After years of research, a vaccine to prevent HIV infection is on the horizon. Get ready to celebrate a future where HIV is nothing more than a distant memory!

So there you have it, folks! The future of vaccines is looking bright – and hilariously bizarre – so roll up your sleeve and get ready for the ride of a lifetime!

World of Exercise

G et ready to strut your stuff in the world of exercise – it's time to walk the walk and talk the talk in the most fun-tastic way possible!

- **Sweat It Out in Style: ** Aim for a fabulously achievable **150** minutes of exercise per week – that's like binge-watching your favorite show, but way healthier!

- **Calorie Crunching Stroll: ** Want to burn some calories? Lace-up those sneakers and hit the pavement – 5 miles a day will have you shedding those pesky calories faster than you can say "snack attack!"

- **Heart-Pumping Party: ** Get your heart rate revved up to 80% of its maximum speed – because who doesn't love a cardio dance party?

- **Fantastic Fitness Fiesta: ** Mix it up with 8-10 different exercises on two nonconsecutive days of the week – it's like a workout buffet for your body!

- **Weight Loss Wonderland: ** Exercise isn't just for fun – it's also your ticket to weight loss and lower blood pressure. Say goodbye to the scale woes and hello to a healthier you!

- **Yoga Yippee: ** Stretch it out with some yoga moves that'll have you feeling Zen AF. Not only does it reduce anxiety and stress, but it also gives your muscles a much-needed boost and keeps you limber well into your golden years.

- **Walk It Out: ** Stick to a walking schedule at the same time every day to keep obesity at bay and your BMI in check. Plus, it's the perfect excuse to show off your fabulous walking shoes!

- **Mind Over Matter: ** Exercise isn't just good for the body – it's also a boon for the brain! Release those feel-good endorphins and soak up some serotonin from the sun to keep those winter blues at bay. Who knew preventing dementia could be so much fun?

- **Longevity Legends: ** Want to activate those longevity genes without breaking a sweat? Just a little jogging here and there, and you'll be turning back the clock without damaging a single cell. It's like the fountain of youth, but with sneakers!

So, there you have it, folks – the ultimate guide to exercising like a boss and living your best life. So, grab those weights, strike a pose, and get ready to slay the day – one fabulous workout at a time!

Exercise- Walk the Walk

Get ready to reap the rewards of exercise – it's like a jackpot of health benefits just waiting for you to cash in! Let's break it down in the most fun-tastic way possible:

- **Heart Happy: ** Say "see ya later" to heart disease with every step you take – exercise is like a shield protecting your ticker from the bad guys!

- **Pressure Drop: ** Knocking out high blood pressure is a breeze when you're busting moves like a dance floor diva. Say goodbye to stress and hello to feeling fabulous!

- **Bones of Steel: ** Who needs superhero powers when you've got exercise? It's like giving your ligaments and bones their own set of magical armor!

- **Stress-Busting Party: ** Stressed out? Exercise is the ultimate stress reliever – it's like hitting the reset button on your brain and saying "adios" to the blues!

- **Snooze Like a Baby:** Get ready for some seriously sweet dreams – exercise is like a lullaby for your body, sending you off to dreamland with a smile on your face.

- **Cancer Crusher:** Fight off those pesky cancers like a superhero in spandex! Exercise is your secret weapon against colon, prostate, and breast cancers – talk about a knockout punch!

- **Fabulous Physique:** Get ready to strut your stuff with confidence – exercise is like a magic wand for your physical appearance, turning heads wherever you go!

- **Ego Boost:** Pump up that self-esteem like a boss – exercise is like a big ol' confidence booster, reminding you just how awesome you truly are!

- **Bone Brigade:** Keep osteoporosis at bay with every squat, jump, and skip – exercise is like a shield for your bones, keeping them strong and sturdy for the long haul!

- **Energy Explosion:** Feeling sluggish? Not anymore! Exercise is like a shot of espresso for your body, giving you a burst of energy that'll make you feel like a superhero in no time!

- **Longevity Legends:** Want to live long and prosper? Lace-up those sneakers and hit the pavement – exercise is like the fountain of youth, keeping you young at heart for years to come!

- **Mind Master:** Keep your brain in tip-top shape with every step you take – exercise is like a secret weapon against dementia, keeping your mind sharp and focused well into old age!

Now, let's talk goals:

- **Medium Fitness Mission:** Aim for 3 miles a day, 5 days a week – that's like strutting your stuff around the block 62 times! Talk about a workout party!

- **High Fitness Fandango:** Ready to kick it up a notch? Shoot for 20 miles a week or more, and get that heart rate pumping to 150 beats per minute – you'll be sweating like a rockstar in no time!

And don't forget your trusty sidekicks Gadgets:

- **Pedometer Pal:** Keep track of your steps like a pro – it's like having a personal cheerleader on your wrist!

- **Elliptical Extravaganza:** Glide your way to fitness on the elliptical machine – it's like dancing on air, minus the awkward moves!

- **Star Stepper:** Take the stairs like a champ – it's like climbing your way to fitness superstardom, one step at a time!

- **Core Crusader:** Get those abs of steel with some killer core exercises – it's like sculpting a masterpiece out of clay, one crunch at a time!

- **Push-Up Power:** Pump up those muscles with some good old-fashioned push-ups – it's like giving your body a high-five for being so awesome!

So there you have it, folks – the ultimate guide to exercise bliss! Now go forth and conquer the world, one fabulous workout at a time!

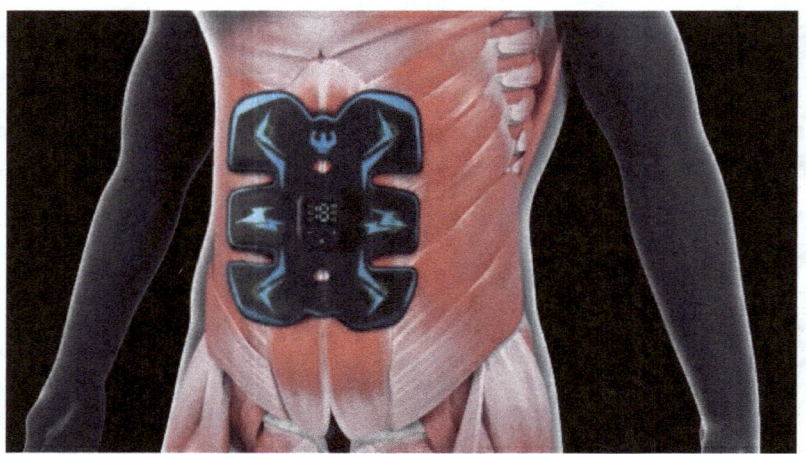

Stimulator for belly fat

Exercise- LDL

L et's dive into the hilarious world of exercise and discover how to keep those LDL levels in check while having a blast:

- **Pump it Up with Anaerobic Awesomeness:** Get ready to rock those muscles with 10 repetitions per set, 3 times a week, on alternate days – it's like a workout party for your muscles!

- **Weightlifting Wonder:** Move over, aerobic exercise – resistance training is here to steal the spotlight! Lift those weights and feel the burn in all the right places.

- **Post-Meal Party:** Who says exercise can't be delicious? Burn off those calories with 30-45 minutes of fun after eating – it's like dessert for your body!

- **Stairway to Fitness:** Climb your way to calorie-burning glory with 2 steps per minute on a stair climber – it's like reaching new heights of fitness while watching your favorite show!

- **Daily Calorie Burn:** Aim to torch at least 300 calories a day with your exercise routine – it's like saying "adios" to those unwanted pounds, one sweat session at a time!

- **Fabulous Five-Step Walking Routine:** Warm up, stretch, walk, cool down, and stretch again – it's like a dance party for your legs, complete with a grand finale stretch!

- **FITT Principles Extravaganza:** Frequency, intensity, time, and type – it's like the ultimate recipe for fitness success! Mix and match your workouts like a pro and watch those gains roll in!

- **BMI Bonanza:** Keep that BMI in check – every unit increase above 19 means more medical costs and drug expenses. Save your pennies and shed those pounds like a boss!

- **Skinny Caliper Shenanigans:** Measure up with a caliper and track your body fat percentage – it's like playing detective with your health, one pinch at a time!

- **Endorphin Euphoria:** Get ready to ride the wave of endorphins – exercise is like a natural high, bringing joy, calmness, and a whole lot of euphoria to your day!

- **Sunshine Serotonin Boost:** Soak up some rays and boost your serotonin levels – it's like sunshine in a bottle, warding off the winter blues and keeping your mood sunny all year long!

- **Vitamin D Delight:** Get your daily dose of sunshine for strong bones and a rock-solid immune system – it's like giving your body a big ol' hug from the sun!

- **Neck Muscle Nonsense:** Strengthen those neck muscles with a simple side-to-side motion – it's like giving your neck a mini workout while waiting for the bus!

- **Shoe Sensation:** Step out in style with good shoes – it's like walking on clouds, with support and comfort every step of the way!

- **Lunchtime Routine:** Keep your meals on schedule for maximum energy and digestion – it's like setting your body's clock to "healthy and happy" mode!

- **Walking Wonder:** Stick to a walking schedule at the same time every day – it's like having a date with fitness, rain or shine!

So there you have it, folks – the ultimate guide to exercise hilarity! Get ready to laugh, sweat, and feel fabulous with every step you take on your fitness journey!

CARVER

DAREBEE BACK WORKOUT © darebee.com

LEVEL I 3 sets **LEVEL II** 4 sets **LEVEL III** 5 sets **REST** up to 2 minutes

20 bridges   **10** V-ups   **20** bridges

10 knee-to-elbows   **20** bridges   **10** side jackknives

INNER
WARRIOR

DAREBEE WORKOUT © darebee.com
Hold each pose for 20 seconds then move on to the next one.
Repeat the sequence again on the other side.

1. warrior I

2. warrior II

3. lunge

4. lunge with twist

5. pigeon pose

6. downward dog

7. bow pose

8. child pose

9. reclining hero

one & one

DAREBEE WORKOUT © darebee.com

1 minute each exercise | **1 minute** rest between each

high knees

jumping jacks

squats

side leg raises

lunges

plank arm raises

plank leg raises

planks with rotations

climbers

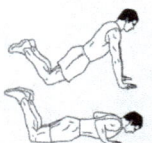

knee push-ups

Wellness Game

Get ready to chill and heat your wellness game with these fun-tastic hot and cold adventures!

1) **Epsom Salt Whirlpool Wonderland:** Dive into a world of relaxation with a heated hydrotherapy massage in an Epsom salt whirlpool – it's like a warm hug for your muscles!

2) **Cayle-Riley Massage Mayhem:** Say goodbye to stress and tension with a massage that'll have you feeling like a noodle in no time – it's like a spa day for your body and soul!

3) **Infrared Sauna Spectacular:** Get ready to sweat it out in style with an infrared sauna session – it's like a hot yoga class without the downward dog! Burn calories, rejuvenate your skin, and chill out like a pro.

4) **Chromotherapy Carnival:** Immerse yourself in a rainbow of colored lights and let the healing vibes wash over you – it's like a disco party for your senses! Say goodbye to inflammation, pain, and stress, and hello to a brighter, happier you.

5) **Tub Ice Pack Adventure:** Take the plunge into icy waters with a tub ice pack – it's like a refreshing dip in the Arctic, with all the invigorating benefits of cold therapy! Say hello to reduce swelling, improve circulation, and a wake-up call for your body.

6) **Cryotherapy Cool Down:** Step into the deep freeze with cryotherapy – it's like a winter wonderland in a box, with temperatures dipping to -110 or -166°F! Say goodbye to sore muscles and hello to a frosty blast of rejuvenation.

And hey, did you know? A sauna bathing study in Finland found that guys who hit the sauna 7 days a week had a twofold drop in heart disease and all-cause mortality compared to those who only sauna once a week. So go ahead, embrace the heat, and chill out like a Finnish rockstar!

So, there you have it, folks – the ultimate guide to hot and cold wellness fun! Whether you're heating up in the sauna or chilling out with a tub ice pack, get ready to feel fabulous from head to toe!

◆ ◆ ◆

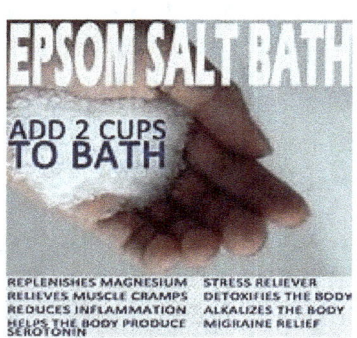

◆ ◆ ◆

◆ ◆ ◆

Infrared Sauna

CHAPTER 6

You are what you eat!
longevity lifestyle diet

Nothing white diet; mainstay food!

here's the information organized into a table format:

Food	**Avoid**	**Limited**	**Best**
No White Flour Products	pasta, white bread, pastries, desserts, bread, bagels	whole grains, Gluten-free, if possible, fruit, poultry, fish, (no skin)	lentil, beans, spaghetti, squash, krait, Organic, fresh low starch. vegetable, lightly

			colored
No Sugar	wheat	courgettis, green (well trimmed)	spinach, kale, collard, cabbage, broccoli, cauliflower, lettuce
No Salt			
No Flour or Corn Tortillas			
No Quick-Cooking Oatmeal	avoid instant		
No Buttered Popcorn	only hot air popped		
No Milk and Cheese		no milk products	unsweetened silk almond milk
No Refined Starch		avoid most commercial breakfast cereals	whole grains, gluten-free, fruit
No Hydrogenated Fat	commercial prepared oils	poultry, fish,	poultry, fish (no skin)

No Commercial Oils		wheat	nuts, avocados, flax seed,
Nuts, Peanuts, Beans			nuts, lentils, beans, spaghetti squash
No Fried Foods			**stir fried.** ostrich, venison, buffalo
No Soft Drinks			
No Alcohol, Caffeine			
No Unfiltered Tap Water			
No Baked or Fried Bread			
No Potatoes, Sweet Potatoes			
No Trans Fats	margarine, shortening		extra virgin olive oil
No High Glycemic Index Foods	avoid quick-cooking instant rice/	commercial cereals, starchy veggies	

	oatmeal		

This table summarizes the "Nothing White Diet" mainstay foods, indicating what to avoid, what to limit, and what are the best choices.

Healthy Eating Guidelines

Here's what you should eat:

Lots of veggies like onions, cucumbers, tomatoes, bell peppers, potatoes, cabbage, garlic, and more.

Eat plenty of legumes such as chickpeas and fava beans.

Snack on nuts like almonds, walnuts, and hazelnuts.

Enjoy whole grains like spelt wheat, semolina, and polenta.

Use olive oil generously, about 3 tablespoons a day.

Sometimes have fish (2-3 times a week), and pick good quality ones like salmon, sardines, or shrimp.

Have eggs and fruits occasionally and treat yourself to dark chocolate in small amounts.

Limit dairy and white meat like chicken.

Avoid red meat, processed meat, added sugar, and processed foods.

Obesity Diet

Check out this wild menu fit for a guy battling with the bulge!

Start your day with some whole-grain toast slathered in jam without all that added sugar. Wash it down with a cup of almond milk, also keeping it sugar-free.

For lunch, dig into a hearty dish of spelt with chicory and carrots. This recipe's got spelt flour, olive oil, cheese, chicory, and carrots. You can swap out the spelt for barley or rice if you fancy.

Grab a snack of a slice of whole-wheat bread and munch on three tiny squares of dark chocolate with at least 70% cacao for that extra health kick.

Come dinnertime, reel in some fresh salmon drizzled with olive oil and served with a side of whole-wheat bread and veggies like cauliflower.

Finish off your feast with some fresh, juicy fruit in season, paired with walnuts or your favorite dried fruit.

Spelt Wheat

Wacky Menu for Golden-Ager

A man who is over 70 years old

Check out this wacky menu for a guy who's a seasoned pro at losing muscle!

For brekkie, munch on whole-grain toast with jam that's not loaded with sugar. Wash it down with a cup of almond milk, also sans sugar.

At lunch, dive into a farro salad jazzed up with lentils and sesame seeds.

Grab a snack of three tiny squares of dark chocolate and munch on 10 dry-roasted hazelnuts.

Come dinner time, slurp up some spaghetti with clams, and don't forget a side of fresh veggies like collard greens.

Wrap up your feast with some fresh, juicy fruit in season, paired with walnuts or your favorite dried fruit.

5 Beginner Tips for Getting Started With the Longevity Diet

Here are some laughable tips to kick-start your adventure into the wacky world of the Longevity Diet:

1. **Tiny Steps for Big Bellies:** If you're struggling to wrap your head around gobbling up heaps of veggies, don't sweat it! Instead, just stick to munching within a 12-hour window. You can stuff your face between 7 am and 7 pm, or push it to 8 am and 8 pm, but whatever you do, don't skip breakfast! We're not monsters here.

2. **Play with Your Food:** Spice up your culinary journey by sampling veggies you've never laid eyes on before! Hit up exotic eateries and take your taste buds on a wild ride. Order some minestrone soup or pasta fagioli at an Italian joint, then try your hand at recreating the magic at home. Feeling adventurous? Dive into Greek cuisine and tackle their horta or sautéed spinach, drenched in lemon juice and olive oil. Trust me, it's a trip worth taking.

3. **Ditch the Junk:** Bid farewell to processed junk like it's a bad ex. Ease into your new relationship with food by dabbling in some cooking escapades. Browse YouTube, TikTok, or cooking websites for inspiration. Challenge yourself to whip up one new recipe each week until you've got a repertoire that would make Gordon Ramsay proud.

4. **Become an Ingredient Detective:** Ever read the back of your favorite processed snacks? Prepare to have your mind blown! Take a leisurely stroll down the grocery aisle and dissect those ingredient lists like a forensic scientist. Once you realize what's lurking in those bags of chips, you'll be sprinting to the kitchen to whip up something fresh and tasty.

5. **Food Detective: The Allergy Edition:** The Longevity Diet

might sound like it's written in stone, but fear not, there's room for customization!

Got pesky food allergies or sensitivities? Don your detective hat and get to the bottom of it. If you're stuck with something like celiac disease, and grains are a no-go, fret not! Seek guidance from a doctor or nutritionist to navigate the maze of dietary restrictions like a pro. After all, we're all about longevity, not living in food fear!

Resources We Love

Behold, the hilarious guide to our top-notch resources for diving headfirst into the quirky world of the Longevity Diet:

Favorite Websites:

1. **ValterLongo.com**: This treasure trove is a foodie's paradise! Get ready to drool over mouthwatering, Italian-inspired recipes that fit snugly into the Longevity Diet playbook. Plus, stay in the loop with cutting-edge research and tips that'll make you the MVP of meal prep.

2. **The Vegan Society**: It's like the Longevity Diet's cool cousin! Dive into a veggie wonderland where plants reign supreme. Whether you're a seasoned herbivore or just curious about veggie magic, this site serves up inspo aplenty. Psst... bookmark it for those fasting-mimicking days!

◆ ◆ ◆

Favorite Online Tool:

- **Monterey Bay Aquarium Seafood Watch**: Fishy business got you scratching your head? Fear not! With this nifty tool, navigating the murky waters of seafood shopping is a breeze. Say goodbye to mercury scares and hello to sustainable fishy feasts!

Favorite Blogs:

1. **ElaVegan**: Brace yourself for a culinary adventure courtesy of our vegan guru! From quinoa lentil burgers to no-bake raspberry cheesecake, these recipes are so good, that you'll forget they're healthy.

2. **The Posh Pescatarian**: Not a seafood fan? Prepare to have your mind blown! Stephanie Harris-Uyidi is the queen of making seafood sing. Say goodbye to fishy fears and hello to mouthwatering pescatarian delights!

Summary:

So, you're ready to embark on a journey toward eternal youth and deliciousness? The Longevity Diet might just be your ticket to a lifetime supply of good vibes and great health. With a focus on veggies, fruits, and all things plant-based, you'll be feeling fresher than a sprout in no time. Just remember, sticking to this diet takes a bit of grit and determination. But fear not, brave soul, for the rewards are as bountiful as a veggie buffet at a unicorn party!

And hey, if the Longevity Diet isn't your jam, there's always the Mediterranean Diet! Picture this: nuts, seeds, olives, and even a splash of red wine and dark chocolate. It's like a culinary vacation to the shores of eternal happiness. So go forth, foodie explorer, and may your taste buds lead you to a life filled with delicious adventures!

Blueberry Chia Seed Pudding- Ela vegan

CHAPTER 7

Dr. G's Food Guide Hilarity

D r. G's Hilarious Guide to Food Survival:

1. Say bye-bye to anything that comes in a suspiciously shiny wrapper or scream "I'm processed!" like a bad actor in a B-movie.

2. Before chowing down, scrub those paws like you're prepping for surgery.

3. And after visiting the loo, because let's face it, you don't want to mix business with pleasure.

4. Clean mitts are a must for whipping up culinary masterpieces. No one wants a side of E. coli with their salad.

5. Hygiene is king. So, bathe regularly, please.

6. When in doubt, douse those hands with enough sanitizer to sterilize a small village.

7. Keep your sheets as clean as your conscience... or at least toss them in hot water once a week.

8. Bedroom clutter is like having an audience for your snoozefest. Keep it out.

9. Brush, floss, repeat. Your dentist will thank you.

10. Share the wisdom of the ages with the next generation. Less sugar equals more brain power, and who doesn't want to crush exams and dominate game night?
11. Embrace your inner rabbit and aim for ten servings of fruits and veggies a day. Go ahead, make Bugs Bunny proud.

12. Ladies, keep it classy with just one drink a day. Fellas, you get two. But let's not turn this into a competition, okay?

13. Breakfast: the most important meal of the day, according to moms everywhere. Fuel up with some fresh fruit or whole grains and tackle the day like a champ.

14. Fresh is best, so hit up your local farm stand or veggie store for seasonal delights. Canned goods need not apply.

Continuing the Food Fiasco:

15. Always leave a little something on your plate to keep the food gods happy. Plus, it's like a surprise snack for your future self.

16. Ditch the dinnerware the size of a small planet. Opt for dainty nine-inch plates and trick your brain into thinking you're feasting like royalty.

17. No second helpings allowed! Unless, of course, you've got a time machine to undo those regrets.

18. Embrace your inner caveman (or woman) with some intermittent fasting. It's like a food rollercoaster for your taste buds.

19. Go green, baby! Vegan is the way to glow, with plant-based goodies that'll have cows breathing a sigh of relief.

20. Junk food? More like a junk mood. Say no to processed goodies that taste like regret and regret alone.

21. Fast food? More like... fast, I'm gonna regret this later. Skip the drive-thru disaster and save yourself from a future food coma.

22. Instant oatmeal? More like instant disappointment. Stick to the real deal, folks.

23. Fresh ingredients are the building blocks of a happy belly. Say goodbye to packaged shortcuts and hello to culinary satisfaction from scratch.

24. Counting calories is like math class for your stomach. But hey, if it keeps you from squeezing into those skinny jeans, count away.

25. Guzzle down that H2-oh yeah! Ten glasses a day will have you hydrated and ready to tackle life's challenges. Plus, bonus points for high pH water – it's like a spa day for your insides.

26. Soft drinks: the nectar of the sugar gods. Except not really. Say no to liquid regret and opt for something that won't turn your insides into a science experiment.

27. Iron, the silent assassin of aging. Keep those free radicals at bay and donate some blood while you're at it. It's like hitting pause on the aging process, one pint at a time.

28. Red meat: delicious, but deadly in excess. Keep it in check or risk playing Russian roulette with your health.

29. Eat less, live more. Cut those calories like you're editing a bad movie, and watch the pounds melt away faster than a popsicle in July.

30. Surround yourself with the svelte squad. Skinny friends are like walking, talking about weight loss motivation. Sorry, chunky pals – guilt by association is real.

More Food Follies to Tickle Your Taste Buds:

31. Fat-free foods: the ultimate diet deception. They may be missing the fat, but they sure ain't missing the calories. Say hello to disappointment and goodbye to your skinny jeans.

32. Say no to smoked and packaged goodies that come with a side of stomach cancer roulette. Bacon, hot dogs, bologna – they may taste like heaven, but they'll send you straight to dietary purgatory.

33. High-fat foods: the villains of the culinary world. Keep those hydrogenated baddies at bay or risk a cholesterol catastrophe.

34. Processed flour: the enemy of the waistline. Wave goodbye to bread, pasta, and anything else that comes in a shade of ghostly white.

35. Sugar sneaks into more places than a Hollywood gossip columnist. Say no to processed tomato sauce and opt for homemade goodness instead.

36. Margarine and shortening: the imposters of the spread world. Butter reigns supreme – accept no substitutes.

37. Start your meal with a broth-based soup and trick your brain into thinking you've won the calorie lottery. It's like a pre-dinner warm-up for your taste buds.

38. Chicken soup: the ultimate cure-all. Forget the doctor – a bowl of homemade goodness will make you feel better in no time.

39. Steamed veggies: the unsung heroes of the dinner table. Keep 'em crisp and colorful, or risk turning your plate into a sad, soggy mess.

40. Gluten-free goodies: the saviors of sensitive stomachs

everywhere. Embrace them like a long-lost friend, and watch your digestion dance with joy.

41. No trans rats allowed! Stick to butter over margarine and watch those pesky rodents of unhealthy fats scurry away in fear. And remember, when in doubt, just say no to anything white – unless it's cauliflower. That stuff's pure magic.

More Food Funnies to Digest:

42. Say "see ya later" to commercial soups and sauces – they're like sneaky sugar and salt ninjas hiding in your pantry. And ketchup? That's just tomato syrup in disguise.

43. Gluten: the ultimate party pooper for your digestive system. Avoid wheat, oats, rye, and barley unless you want your stomach to throw a riot.

44. Fried foods: the guilty pleasure of the culinary world. Resist the siren call of the deep fryer or risk turning into a walking, talking grease trap.

45. Caffeine: the elixir of productivity or the devil's juice, depending on who you ask. Keep it to a sensible two cups a day or risk turning into a jittery mess faster than you can say "Red Bull gives you wings."

46. Coffee curfew: because nothing ruins a good night's sleep like a late-night caffeine buzz. Save the java jitters for the morning and let your dreams be sweet, not caffeinated.

47. Iron-fortified foods: the silent assassins of your digestive system. Avoid them like the plague unless you want your insides to feel like a metal workshop.

48. Commercial cereals: the breakfast of champions... if champions enjoy starting their day with a bowl of sugar-coated regret.

49. Trans fats: the sworn enemies of heart health. Say no to

margarine and shortening and let your arteries breathe a sigh of relief.

50. Quick-cooking rice and oatmeal: the culinary shortcuts that'll leave your taste buds feeling cheated. Embrace the slow and steady cooking process or risk a bland mealtime meltdown.

51. Extra virgin olive oil: the Beyoncé cooking oils. Accept no substitutes and let your dishes shine like Queen Bey on stage.

52. Filtered high-pH water: because hydration is key, and tap water is so last season.

53. High glycemic index foods: the rollercoaster ride for your blood sugar levels. Avoid them like the plague unless you enjoy feeling like a human yo-yo.

54. Lean meat: the protein powerhouse that'll have you feeling like a Greek god... or at least a slightly less guilty carnivore.

55. Ostrich meat: the lean, mean, protein machine of the animal kingdom. Say goodbye to fat and hello to flavor with this exotic delicacy.

56. Butter popcorn: the movie night staple that'll leave your fingers feeling like grease magnets. Stick to air-popped perfection and keep the butter at bay.

57. Toasted rye bread: the unsung hero of the bread aisle. Say goodbye to boring whole grains and hello to a toast-worthy taste sensation.

58. Pre and probiotics: the dynamic duo of gut health. Keep your microbiome happy and watch your stress levels plummet faster than a hot air balloon.

59. Flour or corn tortillas: the carb-loaded culprits of taco night tragedy. Opt for lettuce wraps or risk turning your fiesta into a siesta.

60. Milk and milk products: the lactose-laden landmines of the dairy aisle. Say no to cheese-induced heartburn and opt for alternatives that won't leave you feeling like a bloated balloon.

61. Water before meals: the ultimate appetite suppressant. Guzzle down those eight ounces and watch your portion sizes shrink faster than your patience in a grocery store line.

Let's Keep the Food Fiasco Rolling:

62. Sweet tooth acting up? Say no to Candyland and hello to the dark side – chocolate, that is. Just make sure it's 80% cocoa or higher because life's too short for wimpy chocolate.

63. Buckwheat: the unsung hero of the grain world. Say goodbye to boring old wheat and hello to the buckwheat brigade.

64. Get yourself a veggie steamer and unleash your inner vegetable ninja. Who needs swords when you've got broccoli to conquer?

65. Packaged foods: the sworn enemies of freshness. Say no to canned, processed goodies, and embrace the culinary adventure of cooking from scratch.

66. Protein bars and cereal bars: the imposters of the snack world. Say no to cardboard with a side of disappointment and opt for real food instead.

67. Sports drinks and fruit juices: the liquid sugar bombs of the beverage aisle. Skip the sugar rush and reach for something that won't leave you feeling like a human pinball.

68. Dried fruits: the fiber-packed snacks that'll have you feeling like a human balloon. Say goodbye to fruit-induced flatulence and keep your gas levels in check.

69. Spreads, jams, and preserves: the sticky temptations of breakfast time. Resist the urge to slather and save yourself from

a sugar-induced meltdown.

70. Sauces: the sneaky sugar delivery systems hiding in your pantry. Say goodbye to ketchup's siren call and embrace the naked truth of salad dressing-free salads.

71. Canned fruits: the syrupy pitfalls of convenience. Say no to fruit swimming in sugar water and opt for the real deal instead.

72. Dates, raisins, and watermelon: the sugar-laden snacks that'll have you bouncing off the walls. Say no to the sugar rush and keep your energy levels in check.

73. Whole grains: the rockstars of the pantry. Say goodbye to boring old white bread and hello to the whole-grain dream team.

74. Alcohol: the liquid courage that comes with a side of regret. Say no to boozy binges and embrace a healthier, happier liver.

75. Legumes: the fiber-filled wonders of the plant kingdom. Say hello to beans, peas, and lentils, and watch your digestive system do a happy dance.

76. Nuts and seeds: the snack attacks saviors of the health-conscious. Say goodbye to greasy potato chips and hello to the crunchy goodness of almonds, walnuts, and chia seeds.

Let's Keep the Food Fiasco Rolling:

77. Whole grain rye and bran cereals: the breakfast buddies that'll have your blood sugar snoozing instead of spiking. Say goodbye to breakfast drama and hello to a morning of mellow.

78. Gluten: the sneaky saboteur lurking in your soup bowl and hiding in your hot dog bun. Say no to gluten gremlins and keep your gut feeling groovy.

79. Soy milk: the bloating bandit of the dairy aisle. Skip the soy and keep your hormones in check – unless you enjoy feeling like a human balloon.

80. Chewing gum: the innocent-looking culprit behind your sudden gas attacks. Say no to sorbitol and keep your stomach from staging a bloated rebellion.

81. Almond milk: the dairy alternative with a dangerous sidekick named carrageenan. Stick to brands like Trader Joe's and Whole Foods 365 and avoid turning your morning latte into a stomachache symphony.

82. Extra virgin olive oil: the superhero of the kitchen. Say goodbye to oil overload and hello to the perfect balance of healthy fats.

83. Mushrooms: the cancer-fighting superheroes hiding in your produce aisle. Say hello to your new favorite fungus and wave goodbye to cancer worries.

84. Modified food by ramid: the mysterious stranger of the grocery aisle. Say no to Frankenfoods and keep your pantry free from genetically modified monsters.

85. Sea vegetables: the oceanic wonders of the superfood world. Dive into the deep end with seaweed snacks and watch your health soar to new heights.

86. Artificial sweeteners: the chemical chaos-makers of the sugar substitute scene. Say no to saccharin, aspartame, and sucralose, and opt for the natural sweetness of stevia instead.

87. Calorie reduction: the secret sauce of weight loss success. Say goodbye to excess calories and hello to a slimmer, happier you – just don't forget to save room for dessert (in moderation, of course).

Well, well, well, if it isn't Mr. Grumpy's Grump-tastic Guide to Food Fussiness:

88. Organic foods: because apparently, Mother Nature needs a fancy USDA seal of approval to be considered legit. And don't

forget to wash your produce like you're giving it a spa day –
warm water only, please.

89. Fresh raw vegetables with red wine vinegar and extra virgin
olive oil: because apparently, salad dressing needs an invitation
to the party. Hold the croutons, we're going full grump mode.

90. Knox plain gelatin in fruit juice or vegetable juice: because
apparently, regular old juice isn't exciting enough on its own.
Just add gelatin and voila, instant party in your mouth. Or not.

91. Vegetables: the above-ground vs. below-ground showdown.
Where veggies grow is a matter of high-stakes importance. Keep
it non-starchy, people – our grumpy guts demand it.

92. Sunflower seeds: the unsung heroes of salads and fish.
Regular old toppings just won't cut it anymore. Sprinkle those
seeds and watch your taste buds... Well, I don't care that much.

93. Monticello: because apparently, Thomas Jefferson's Garden
is the ultimate authority on herbs and veggies. Forget about
modern farming techniques – let's go full 18th-century grump.

94. Food ninja appliances: because apparently, regular kitchen
gadgets just don't have enough attitude. Ninja up your cooking
routine and watch your culinary skills... well, still not impress
anyone.
95. Xylitol gum: because apparently, regular gum just doesn't
have that bitter aftertaste we all crave. Say hello to ricochet and
power Brite – the gums that'll leave your taste buds feeling...
well, bitter.

96. Kombucha: because apparently, regular old drinks just don't
have enough fermented funk. Drink out of glass bottles only,
because apparently, plastic is the devil's drinkware.

97. Microwaves: because apparently, convenience is overrated.
Say no to plastic containers and hello to glass – because
apparently, microwaving plastic will turn your leftovers into

toxic waste.

98. Glucose-producing foods: because apparently, carbohydrates are the enemy of the people. Say goodbye to bread and dairy – because apparently, wheat and dairy are the devil's snacks.

99. Ideal body fat: because apparently, we all need another reason to stress about our waistlines. Say hello to the grumpy gut brigade – because apparently, anything over 20% body fat is a one-way ticket to the grump zone.

100. Red meat: because apparently, carnivores are just asking for heart disease. Say goodbye to steak – because apparently, even the thought of a juicy burger will give you heart palpitations.

101. Intermittent fasting: because apparently, skipping meals is the new black. Say hello to the hangry brigade – because apparently, starving yourself is the key to eternal grumpiness.

102. LDL levels: because apparently, keeping your cholesterol in check is the key to a heart attack-free life. Say hello to the cholesterol police – because apparently, LDL levels are the gatekeepers of cardiac catastrophe.
103. Food rules: because apparently, eating is now a complicated, multi-step process. Say hello to the grumpy diet plan – because apparently, simplicity is for suckers.

104. Alkaline water: because apparently, regular old H2O just doesn't cut it anymore. Say hello to the pH police – because apparently, your blood needs to be as alkaline as your attitude.

105. Lentils and beans: because apparently, regular old fiber just isn't enough. Say hello to the gas brigade – because apparently, bloating and flatulence are the price you pay for slow-carb digestion.

106. Alkaline water machines: because apparently, regular old filters just aren't fancy enough. Say hello to the water purifier – because apparently, ultraviolet light is the secret to hydration.

107. Carbohydrates: because apparently, regular old carbs just don't have enough fiber. Say hello to the bloating brigade – because apparently, flatulence is the price you pay for carb consumption.

108. Freshly squeezed organic vegetable juice: because apparently, regular old juice just doesn't have enough vitamins. Say hello to the grumpy gut cleanse – because apparently, vegetables are the answer to all of life's problems.

109. Invigor-8: because apparently, regular old supplements just don't have enough attitude. Say hello to the grumpy gut health pack – because apparently, raw coconut oil is the key to happiness.

Thomas Jefferson's Garden

Let's Get Burpy with Food Facts:

110. Grilled or pan-fried meat: because apparently, nothing says "yum" like a side of heterocyclic amines with your steak. Say hello to potential harm and wave goodbye to your appetite – those burps might be the least of your worries.

111. Organic fruit skin: the superhero cape of the produce aisle. Say hello to phytochemical protection and wave goodbye to cancer worries – those burps just got a whole lot healthier.

112. Filtered tap water: because apparently, bottled water just doesn't have enough flavor. Say hello to the taste of victory and wave goodbye to plastic waste – those burps are bubbling with environmental pride.

113. Coffee: the bean juice of memory and attention. Say hello to temporary brain boosts and wave goodbye to midday slumps – those burps are like a caffeinated symphony for your frontal lobe.

114. Soft drinks: the tooth decay villains of the beverage aisle. Say hello to irreversible damage and wave goodbye to pearly whites – those burps might be the least of your dental worries.

115. Age spots: the unwelcome guests on your skin's party. Say hello to elevated glucose levels and wave goodbye to youthful glow – those burps are like a reminder of your sugary sins.

116. Cayenne pepper: the fiery appetite extinguisher. Say hello to spicy satisfaction and wave goodbye to overeating – those burps are like a fiery finale to your mealtime drama.

117. Bouillon soup: the appetite assassin of the soup world. Say hello to broth-based bliss and wave goodbye to hunger pangs – those burps are like a savory symphony in your stomach.

118. Nuts: the pre-meal snack saviors. Say hello to crunchy cravings and wave goodbye to overindulgence – those burps are like a nutty nod to portion control.

119. Shrimp cocktail fork: the tiny titan of portion control. Say hello to petite plates and wave goodbye to oversized servings – those burps are like a delicate dance on your taste buds.

120. Sunflower seeds: the salad and fish sidekick. Say hello

to crunchy companionship and wave goodbye to bland bites – those burps are like a symphony of flavor in your mouth.

121. Breakfast: the morning masterpiece. Say hello to salmon sensations and wave goodbye to boring breakfasts – those burps are like a wake-up call for your taste buds.

122. Fruit-infused water: the hydration sensation. Say hello to refreshing sips and wave goodbye to plain old H2O – those burps are like a fruity fiesta in your belly.

123. Gum disease: the heart hazard hiding in your mouth. Say hello to oral hygiene and wave goodbye to cardiac concerns – those burps are like a clean bill of dental health.

124. Nutrition bars: the sneaky sweets in disguise. Say hello to candy-covered confusion and wave goodbye to false health claims – those burps are like a warning sign from your gut.

125. Hormones: the appetite architects of your body. Say hello to satiety signals and wave goodbye to mindless munching – those burps are like a symphony of hunger control.

126. Cranberry juice: the urinary tract superhero. Say hello to infection prevention and wave goodbye to bathroom battles – those burps are like a victory lap for your bladder.

127. Green tea: the cancer-fighting champion. Say hello to antioxidant armor and wave goodbye to free radical fears – those burps are like a toast to your health. Cheers!

Oh, what a shocker! Who would've thought that stuffing your face with all those ultra-processed goodies could lead to an early checkout from Hotel Hypertension? I mean, who needs fresh, wholesome ingredients when you can just pop open a bag of chemical-laden snacks and roll the dice on your health, right?

And hold the phone – cranberry juice to the rescue! Apparently, a daily dose of tart goodness is all it takes to keep those pesky urinary tract infections at bay. Move over, antibiotics – there's a new sheriff in town, and it's wearing a cape made of cranberries.

Because, you know, nothing says "living on the edge" like chugging down sugary, processed concoctions and hoping for the best. But hey, if it means dodging a UTI or two, I guess we can all raise a glass to the power of cranberry juice – just make sure it's not mixed with anything too processed. We wouldn't want to cancel out all those antioxidant benefits with a side of sodium overload now, would we?

Let's Stuff That Stomach
Full of Food Fun:

128. Broccoli and cauliflower: the dynamic duo of cancer prevention. Say hello to veggie victory and wave goodbye to prostate problems – those stomach-stuffing bites are like a green guardian for your manhood.

129. Cholesterol-lowering drugs: the wonder pills of cancer prevention. Say hello to pharmaceutical protection and wave goodbye to colon concerns – those stomach-stuffing doses are like a medical shield for your gut.

130. Free-range chicken and eggs: the farm-fresh delights of ethical eating. Say hello to clucking good choices and wave goodbye to factory-farmed regrets – those stomach-stuffing bites are like a poultry paradise for your palate.

131. Goat cheese: the creamy delight of the cheese world. Say hello to goaty goodness and wave goodbye to lactose intolerance – those stomach-stuffing nibbles are like a cheesy dream come true.

132. MSG: the sneaky flavor enhancer hiding in your pantry. Say hello to natural flavors and wave goodbye to mystery ingredients – those stomach-stuffing bites are like a culinary adventure without additives.

133. Commercially prepared oils: the greasy villains of the kitchen. Say hello to homemade goodness and wave goodbye to deep-fried disasters – those stomach-stuffing meals are like a chef's kiss of approval.

134. Fried foods: the guilty pleasures of the fast-food world. Say hello to oven-baked alternatives and wave goodbye to grease stains – those stomach-stuffing bites are like a crispy crunch without regret.

135. Unfiltered water: the murky menace of hydration. Say hello to alkaline filtration and wave goodbye to tap water troubles – those stomach-stuffing sips are like a refreshing oasis in a desert of dehydration.

136. Artificial sweeteners: the chemical concoctions of sugar-free living. Say hello to natural sweetness and wave goodbye to saccharin sorrows – those stomach-stuffing treats are like a guilt-free indulgence for your sweet tooth.

CHAPTER 8

Piggy Perils

A h, behold the horrors of the porkocalypse! Let's dive into the grisly tale of why piggy products should be banished from your plate like an ancient curse.

Picture this: a world where every juicy bacon strip is a harbinger of doom, where every succulent pork chop is a portal to gastrointestinal Armageddon! Pork, dear friends, is the devil's charcuterie, a festering cesspool of salt, fat, and preservatives. It's like dining in a horror movie, where the main course is a jump scare for your colon!

But fear not, for Tufts University, has unveiled the porky apocalypse statistics: a whopping 29% chance of unleashing the wrath of colorectal cancer upon any man foolish enough to indulge in these processed porcine delights. And let's not forget the delightful possibility of trichinosis, a charming parasitic infection waiting to turn your insides into a twisted carnival of pain and regret.

Oh, but cooking your pork to a crisp, you say? A futile endeavor, my friend! Freezing is but a feeble attempt to ward off the insidious larvae of Trichinella, lurking in every uncooked morsel. These little critters thrive on garbage, scraps, and even the occasional rodent, just waiting to invade your intestinal fortress and turn your muscles into their playground.

And the horrors don't end there! Ever heard of the pork tapeworm? Oh, it's a delightful creature, transmitted by pigs feasting on human excrement. Yes, you read that right. Eating uncooked pork is like playing Russian roulette with tapeworms, or worse, cysticercosis, a nightmarish condition where parasites set up shop in your brain, leading to a symphony of neurological chaos.

But wait, there's more! Did you know that 20 million souls fall victim to cysticercoids each year, with 50,000 meeting their demise in a tragic ballet of agony? And all because some folks just can't resist the siren song of a sizzling pork sausage.

So, my friends, heed this cautionary tale: beware the swine, for they are the harbingers of doom, the architects of intestinal Armageddon! Cast aside your hot dogs, your bologna, and your ham, and embrace a life free from the tyranny of processed pig flesh.

Your colon will thank you.

Conserve Water for Life

Ah, behold the majestic dance of the H2O molecule, the elixir of life, the liquid gold that sustains us all! Let us embark on a whimsical journey through the whims and wonders of water, where survival is but a splash away!

Now, picture this: you, dear reader, a mere mortal weighing in at 85 kilograms, wandering through the desert of dehydration, your parched throat crying out for salvation!

Fear not, for the water deities have bestowed upon us the sacred formula for hydration: 1500 milliliters for the first 20 kilograms, then an additional 20 milliliters for every kilogram thereafter. It's like a mathematical incantation to summon the rains of rejuvenation!

So, let's crunch those numbers, shall we? For our stout 85-kilogram fellow, that's 1500 milliliters for the initial 20 kilograms, plus 20 milliliters multiplied by the remaining 65 kilograms, resulting in a grand total of 2,800 milliliters per day. Or, if you prefer simplicity, a nifty 35 milliliters per kilogram.

But wait, there's more! Let's not forget our trusty BMI, the body mass index, a mystical calculation involving your weight in kilograms divided by your height in meters squared. It's like a mathematical riddle whispered by the ancient sages of the scale!

So, if our intrepid adventurer weighs in at 85 kilograms and stands at a lofty 1.82 meters squared (or approximately 338 meters if we're feeling particularly whimsical), their BMI would be...well, let's leave that to the mathematicians, shall we?

But fret not, for in the grand tapestry of life, water is the thread that binds us all, the liquid embrace that cradles us in its aqueous arms. So drink up, dear friends, and let the waters of hydration wash over you like a gentle tide of vitality!

Dive into Health: The Importance and Benefits of Water

Water is not just a thirst-quenching liquid; it's the elixir of life, the cornerstone of our existence, and the unsung hero of our well-being. In this article, we delve into the depths of the importance and myriad benefits of water for our bodies and minds.

1. Essential for Survival:

Let's start with the basics: water is vital for our survival. Our bodies are composed of approximately 60% water, and every cell, tissue, and organ rely on it to function properly. From regulating body temperature to lubricating joints and flushing out toxins, water plays a crucial role in keeping us alive and kicking.

2. Hydration Station:

Staying properly hydrated is like giving your body a refreshing spa day from the inside out. It helps maintain the balance of bodily fluids, which in turn supports digestion, nutrient absorption, and circulation. Dehydration, on the other hand, can lead to a myriad of unpleasant symptoms like headaches, fatigue, and even confusion. So, sip on that water bottle like it's your fountain of youth!

3. Weight Management Wonder:

For those on a quest to shed a few pounds, water is your silent ally. Drinking water before meals can help curb appetite, making you feel fuller and potentially reducing calorie intake. Plus, water has zero calories, making it the ultimate guilt-free beverage choice for those watching their waistlines.

4. Skin Salvation:

Forget expensive skincare products; water is nature's best-kept beauty secret. Hydrated skin is happy skin, boasting a natural glow and elasticity that can't be bought in a bottle. Plus, staying hydrated helps flush out toxins and impurities, giving you that coveted "I woke up like this" radiance.

5. Mental Clarity Champion:

Feeling a bit foggy-headed? Reach for a glass of water before you reach for that extra cup of coffee. Studies have shown that even mild dehydration can impair cognitive function, leaving you feeling sluggish and unfocused. So, keep that brain hydrated and firing on all cylinders with regular sips of Agua.

6. Exercise Enhancer:

Whether you're hitting the gym or going for a leisurely stroll, proper hydration is key to maximizing your workout performance. Water helps regulate body temperature, prevent cramps, and maintain electrolyte balance, ensuring you can go the distance without hitting the dreaded wall of fatigue.

In conclusion, water isn't just a beverage; it's a lifeline, a wellness warrior, and a fountain of health. So, drink up, my friends, and toast to the liquid gold that keeps us thriving from the inside out. Here's to water: our greatest ally in the quest for vitality and well-being!

Youthful Diet Tips

Alrighty kiddos, listen up! Imagine your body is like a super-duper cool recipe – you are what you eat! But hey, just eating healthy snacks won't magically stop you from turning into a wrinkly raisin.

Nope, not gonna happen! Let me spill the beans on some wild stuff scientists are cooking up to fight off the old-age boogie monster.

So, picture this: scientists wake up every day to a flood of messages, like emails from people all over, begging for secrets to staying forever young. Some even ask if we can make their pet hamster live forever – talk about ambitions!

But hold your ice cream cones, some stories are sadder than a dropped popsicle. Like this one time, a dude wanted to donate to science to honor his mom who suffered a bunch from getting old. Then there's this lady, desperate to save her dad from memory blips, willing to spend all her allowance to get him into a study.

Now, here's the scoop: while there's hope on the horizon, battling old age right now is like trying to play tag with a ghost – doctors are still figuring out the rules. But fear not, even without fancy gizmos, we can still tickle our longevity genes and kickstart our journey to eternal youth!

Think about places where folks live forever, like Okinawa, Japan; Nicoya, Costa Rica; and Sardinia, Italy – aka the Cool Kids Club of Centenarians. They munch on veggies like they're candy and go light on the meat and sugar. It's like their secret sauce for staying spry!

Now, listen close, pals. While there's no one-size-fits-all diet for us humans (we're all special snowflakes, after all), there are some general rules to follow: more veggies, less meat, and fresh food over processed junk. Easy peasy, right?

But here's the juicy bit: just like how we beat those old-timey diseases like the flu and tummy troubles, we can totally give aging a run for its money. No more thinking of old age as some inevitable monster lurking in the closet – we can kick its wrinkly butt!

Sure, we might not have a magic potion to live forever (yet!), but by making smart choices – starting with what we chomp on – we can dance through life like eternal kiddos at a never-ending birthday bash. So, remember, pals: eat your veggies, skip the junk, and let's party on, forever young!

Intermittent Fasting, or IF

Alright, buckle up for a wild ride through the land of fasting!

So, you know how during Ramadan, Muslims fast during the day? Well, some super-duper dedicated folks fast twice a week too, channeling their inner fasting ninja vibes on Mondays and Thursdays. And guess what? It's not just some cool trend – it's a legit way to level up your health game!

Now, after diving headfirst into the deep sea of aging research for, like, a gazillion years, if there's one golden nugget of wisdom I can toss your way, it's this: Eat. Less.

I know, mind-blowing, right? But hey, this idea isn't exactly fresh out of the oven. Ancient smarty-pants like Hipocacs and Christian monk Evarist Porticus were all about cutting back on the chow for a healthier, longer life. They were onto something, my friends!

Now, before you start picturing yourself wasting away like a hungry skeleton, hold up! We're not talking starvation mode here. Nope, we're talking about a little thing called fasting – giving your body a break from the non-stop buffet of food we're used to in our comfy, cozy world.

The whole fasting craze got a reboot during World War I when science pals like Lafayette Mendel and Thomas Osborne discovered that rats who didn't pig out on grub lived way longer lives. Yup, rats beating the clock – who knew?

Then along came Clive McCay, the OG of rodent diets, feeding rats cardboard-like cellulose and watching them strut their stuff for longer. And guess what? It worked! Those rats were living their best rodent lives, no joke!

Fast forward through decades of studies and experiments, and it turns out that cutting back on calories is like hitting the pause button on aging for all sorts of critters – from mice to fruit flies to even those teeny-tiny yeast cells. Yup, yeast cells dodging the wrinkles – it's a wild world out there!

But here's the kicker: it's not just about starving yourself into oblivion. Nah, it's about finding that sweet spot – giving your body just enough fuel to keep chugging along, but not too much to send it into a food coma. It's like telling your genes, "Hey, let's stay young and spry forever, deal?"

Now, before you start raiding the fridge in a panic, chillax! Testing this fasting magic on humans is a tad trickier than with our furry friends. I mean, nobody wants to sign up for a starvation boot camp, right?

So, while we might not have all the answers just yet, remember this: less can be more when it comes to chowing down. So, next time you're eyeing that second slice of cake, think twice – your future self might just thank you for it!

Calorie Restriction Gig

Alright, let's break down this whole calorie restriction gig – in layperson's terms!

Picture this: scientists wanna test if eating less can keep

us spry and youthful. But here's the twist – getting folks to chow down less in the name of science ain't as easy as pie. People love their snacks too much!

So, this Duke University gang tried to rope in some brave souls for a food experiment. They asked these adults to cut back on noms by a whopping 25%. But, surprise surprise, folks only managed to slash about 12% of their food intake over two years.

Still, even that little change led to big improvements in their health – like a magical potion for staying young!

Now, imagine chatting with these calorie-cutting pros, Meredith and Paul, who are like the Batman and Robin of eating less. They're all about slashing their food intake by a hefty chunk and feeling awesome about it. Sure, they admit feeling hungry at first, but hey, you get used to it – just like wearing skinny jeans!

And get this – they even have their own style, rocking turtlenecks in summer 'cause, you know, less fat means needing more layers. But don't let their lean frames fool you!

Paul, in his late 60s, was still crushing it as a CEO and chess champ. Plus, his blood work screamed "young and fabulous" on his 70th birthday – just like those calorie-cutting rats!

Now, here's where things get wild. Remember those monkeys chilling on their reduced-calorie diets? Well, they were basically living their best monkey lives, hitting ages way beyond what's normal in monkey years. It's like the fountain of youth but for monkeys!

And it's not just monkeys – even mice are getting in on the action, living longer and healthier on the calorie-cutting train. But here's the kicker: starting early is key. Think of it like jumping on the anti-aging bandwagon before the wrinkles start setting in – gotta beat 'em to the punch!

Now, before you start tossing out all your snacks, know this:

calorie restriction isn't for everyone. It's like committing to a hardcore diet plan that might make you feel like you're living in a food jail. But hey, the perks? It's not just about adding more candles to your birthday cake – it's about staying vibrant and kicking butt in the game of life!

And guess what? Science is cooking up other ways to tap into those same benefits without feeling like you're missing out on all the tasty treats. So, whether you're munching on celery sticks or devouring a slice of pizza, remember – staying young ain't just about counting calories, it's about living your best, most vibrant life, one bite at a time!

◆ ◆ ◆

Fasting and Health

Alright, let's decode this fancy talk about fasting and health – in plain ol' English!

So, you know how when you first try something new, like a scary roller coaster, it feels super stressful? But after a while, you kinda get used to it, and it's not as scary anymore. Well, that's sorta like what happens with fasting!

Now, let's talk about intermittent fasting, or IF for short. It's like giving your tummy a little break from eating now and then. But guess what? This isn't some newfangled idea – scientists have been poking around this concept for ages!

Back in the day, researchers did some experiments on rats (poor critters!), making them skip meals every third day. And guess what? Those hungry rats ended up living way longer than their snack-loving pals! So, basically, skipping meals isn't just about giving your body a chill day – it might do some cool stuff inside your cells!

Now, fast forward to today, and human studies are backing up what those rat experiments hinted at. People who try out this "fasting mimicking" diet – where they eat regular food most of the time but cut back big time on calories for a few days each month – end up shedding some pounds, lowering their blood pressure, and even kicking that pesky hormone called IGF-1 to the curb.

IGF-1 might sound like some weird sci-fi thing, but it's a big deal when it comes to aging. Turns out, lower levels of this hormone are linked to a longer, healthier life. So, basically, by giving your body a little fasting break now and then, you might be giving yourself a ticket to the fountain of youth – without all the wrinkles!

And get this – some lucky folks are born with genes that put them in a sort of fasting mode, no matter what they eat! Yup, you heard that right. Some people can live it up, munching on fried foods, puffing on cigarettes, and sipping drinks, and still make it to 100 with a smile on their faces!

So, next time you hear someone talking about fasting or skipping meals, don't roll your eyes – they might just be onto something! After all, who wouldn't want to live a long, happy life without giving up all the tasty treats?

◆ ◆ ◆

Longevity Village in Bama, Guangxi, China

A lrighty, kiddos, let's dive into the fascinating world of living longer and healthier!

So, you know how some lucky folks seem to hit the genetic jackpot and live super long lives? Well, the good news is, we can all give ourselves a little boost, even if we didn't win the genetic lottery!

How? By tinkering with something called the epigenome – it's like the control panel for our genes, and we can actually tweak it by how we live!

Now, let's talk about a magical place called Blue Zones – like Ikaria, **Greece**, where people seem to have cracked the code to live forever! Turns out, lots of them follow this cool thing called fasting, where they take breaks from eating certain foods.

Like, imagine going without meat, dairy, or eggs for some days – sounds tough, right? But for these Greek pals, it's just part of their routine, especially before important religious stuff like Holy Communion!

And it's not just Greece – there's a place in **China** where folks skip breakfast and only chow down on a big meal later in the day. It's like they're playing hide-and-seek with food for over sixteen hours each day – talk about a hunger game!

Now, here's the fun part – there are tons of ways we can try out this fasting thing, and it's not all about feeling hungry all the time! You can try skipping breakfast and having a late lunch, or maybe cut back on calories for a couple of days a week – it's like a food adventure!

And get this – not only does it help us live longer, but it also saves us some moolah! Plus, if you're not used to stuffing your face all the time, fasting might be a piece of cake (or should I say, a slice of carrot?)!

Now, full disclosure – I've tried eating less, but let's face it, food is just too yummy to resist! Sometimes I accidentally forget to eat, though – oopsie!

But wait, there's more! It's not just about how much we eat, but what we eat too! So, while we're on this journey to living our best, longest lives, let's remember to munch on the good stuff – like veggies, fruits, and all those yummy treats that keep our bodies happy and healthy!

So, there you have it, folks – whether you're fasting like a Greek god or just enjoying a tasty salad, here's to living our longest, happiest lives ever!

Bama, Guangxi, China

◆ ◆ ◆

Superheroes of our bodies
– amino acids!

Alright, kiddos, let's talk about the superheroes of our bodies – amino acids! These little guys are like the Lego blocks that build all the cool stuff in our bodies, like muscles and enzymes. But here's the catch – our bodies can't make all of them on their own, so we gotta get them from our food.

Now, you might think, "Hey, why not just munch on meat? It's packed with all those essential amino acids, right?" Well, yes, but there's a twist – meat can be a bit of a troublemaker. It can clog up our arteries and even play a role in some not-so-fun diseases like cancer. Yikes!

But fear not, young adventurers! There's a way to get our amino acid fixed without the meaty drama. Cue the plants! Veggies, beans, nuts – they're like the rockstars of plant-based protein. They might not have all the amino acids in one go, but together, they make a super team that keeps us healthy and strong.

Now, here's where it gets cool – turns out, when we give our bodies a break from certain amino acids, it's like hitting the refresh button! Our cells go into superhero mode, cleaning up all the gunk and keeping us in tip-top shape.

So, what's the secret sauce to living long and strong? It's a mix of eating smart, moving our bodies, and giving our genes a little nudge now and then. Think of it like taking our bodies out for a spin – we gotta rev up those engines and let our superhero genes do their thing!

So, next time you're munching on a delicious veggie burger or slurping up a smoothie, remember – you're not just feeding your tummy, you're fueling your superhero genes for a lifetime of adventures!

CHAPTER 9

Get Moving with Exercise

L et's Get Moving!

Hey there, kids! Time to dive into the wild world of exercise! Get ready to discover why exercise isn't just about making your heart and lungs happy—it's also about making your cells throw a party!

So, you might think exercise is all about pumping blood faster and getting those muscles jacked up. Well, sure, that's part of it. But guess what? Real magic happens at the cellular level. Yep, we're talking about stuff so tiny you need a microscope to see it!

Picture this: inside your body, there are these things called telomeres. They're like little caps on the ends of your chromosomes, and they're super important for keeping your cells healthy. And guess what? When you exercise, those telomeres get longer! It's like they're doing the limbo dance and stretching out to stay young and spry.

Now, why does exercise make your telomeres do the happy dance? Well, it's all about survival mode, baby! When you work those muscles, your body kicks into gear, revving up energy production and building extra blood vessels. It's like giving your cells a pep talk: "Hey, you've got this! Stay strong and keep ticking!"

But wait, there's more! There are these cool genes in your body called longevity regulators, and they love it when you exercise. They're like the superheroes of your cells, swooping in to protect your telomeres and keep everything running smoothly.

Now, I know what you're thinking: "Do I have to spend hours at the gym to get all these awesome benefits?" Nope! Even just a quick jog around the block or a game of tag with your friends can work wonders. Plus, studies show that even a little bit of exercise each day can add years to your life. So, lace up those sneakers and get ready to feel the power of playtime!

But here's the kicker: to really unleash your body's superhero genes, you gotta kick it into high gear. That means getting sweaty, breathing hard, and feeling the burn! It's all about pushing yourself just enough to make your body sit up and take notice.

So, next time you're tempted to veg out on the couch, remember: your cells are counting on you to get up and get moving! Whether it's a brisk run or a dance party in your living room, every little bit counts. So, let's sweat it out and show those telomeres who's boss!

Roll with the Epigenome!

Alrighty, folks, let's talk about the wild ride of aging and how not to rock the boat too much! You see, a little bit of stress can be good for our bodies, especially when it comes to keeping those longevity genes happy and healthy.

Think of it like this: inside each of us, there's a whole team of DNA repair crews and epigenetic superheroes. When things get rough, they spring into action, fixing up any damage and keeping our bodies ticking along. But here's the catch: they can only handle so much chaos before things start to go haywire.

Now, let's talk about the usual suspects that can wreak havoc on our epigenome. First up, we've got cigarettes. Yep, those little sticks of death are like a wrecking ball for your DNA. They cause so much damage that our repair crews are working overtime just to keep up. So, if you're a smoker, consider kicking the habit for the sake of your poor, overworked cells!

But even if you're not puffing away on cigarettes, there's still plenty of trouble out there. From pollution to chemicals in our food and drinks, it seems like everywhere we turn, there's something trying to mess with our DNA. It's like we're living in a giant game of dodgeball, and the balls are made of toxic sludge!

And let's not forget about radiation, whether it's from the sun or those pesky airport scanners. Sure, they might not turn you into a superhero, but they can definitely cause some DNA damage if you're not careful.

But here's the thing: we can't avoid all the bumps and bruises that life throws our way. Even the act of simply being alive can cause DNA damage. It's like our bodies are constantly playing a game of tag with time, trying to outrun the aging process.

But fear not, my friends! Even if you're not a spring chicken anymore, there's still hope. We may not be able to turn back the clock entirely, but with a little help from science and maybe a few lifestyle changes, we can slow it down.

So, whether you're a teenager or a golden oldie, it's never too late to start taking care of your epigenome. So let's roll up our sleeves, put on our superhero capes, and show those pesky DNA gremlins who are boss!

◆ ◆ ◆

Forever Young Pill

Tiny Enzymes Work Magic.

Once upon a time, deep down in the teeny-tiny world inside our bodies, there was a magical pill – let's call it the "Forever Young Pill." This pill was like a superhero, fighting the evil forces of aging!

Imagine if we could shrink down, down, down, past all the cells and tiny bits inside them, to a place where things are so small, it's like a whole other universe. There, tiny bits called enzymes, which are like little Pac-Men made of proteins, do amazing things. They snatch up molecules and transform them, like turning sugar into energy or making new proteins.

But wait, don't let your eyes glaze over! I know it sounds like a snooze-fest, but trust me, it's super cool! These enzymes are like tiny, hyper-fast workers in a bustling city, constantly zipping around and making magic happen. Without them, we'd be in big trouble – like a car without gas or a phone without battery!

Picture this: inside each of your cells, there are thousands of these enzymes partying it up in a salty sea. It's like a wild dance party where molecules crash into each other at crazy speeds. But in all this chaos, there's a beautiful order that keeps life ticking. If everything stopped suddenly, we'd be goners faster than you can say "pickle juice."

But here's the best part: scientists are like wizards, discovering ways to tweak these enzymes with special potions we call medicines. And guess what? These potions could help us stay sprightly and youthful for way longer!

So, next time you toast to life, remember to raise a glass to these unsung heroes – the enzymes! And who knows, with a bit of science magic, maybe one day we'll crack the code to stay forever young. But until then, let's enjoy the journey, one tiny step at a time!

Flower Galega Officinalis - Metformin

Alright, let's dive into the world of Metformin, the superhero pill that's like pennies for prolonged vitality!

So, picture this: there's this cute little flower called Galega officinalis, but it goes by some weird names like goat's rue or French lilac. Despite its odd names, it's been used as medicine for ages. Turns out, it's packed with a chemical called guanidine, which sounds fancy but basically helps with sugar stuff in our bodies.

Back in the day, doctors were like, "Hey, let's try using guanidine to help people with diabetes control their blood sugar!" And guess what? It worked! Fast forward to now, and we have Metformin, the superstar drug derived from this flower, helping millions of folks worldwide.

Now, diabetes comes in two flavors: type 1, where the body doesn't make enough insulin, and type 2, where the body doesn't listen to the insulin it makes. Metformin swoops in to save the day for type 2 folks by making their cells pay attention to insulin again and keeping sugar levels in check.

But here's the twist: Metformin isn't just about diabetes anymore. Researchers noticed something cool – people taking Metformin seemed to live healthier lives, almost like they found the fountain of youth in a pill! Studies in mice even showed they lived longer and stayed healthier, like little mouse superheroes.

Plus, Metformin seems to have some ninja moves against cancer too! It's like having a shield against those nasty cells gone rogue. And get this, it's not just about cancer – Metformin might help fend off other baddies like dementia, heart problems, and even depression.

But wait, there's more! Metformin doesn't just tackle one thing – it's like multitasking for your health. It boosts enzymes that keep our cells young and helps our bodies fight aging itself! Imagine having a sidekick that helps you stay strong and sharp as you grow older.

Now, the big question is, could Metformin be the key to unlocking longer, healthier lives for everyone? Researchers are working hard to find out. If they crack the code and prove that Metformin can really slow down aging, it could change the game for how we think about getting older.

So, here's to Metformin – the little pill with big dreams of helping us all live longer, healthier lives! Who knew a flower could hold the secret to a supercharged life?

◆ ◆ ◆

Secrets of Aging

Once upon a time, in a lab far, far away, scientists were on a quest to unlock the secrets of aging. They were like detectives searching for clues in the tiny world of cells.

But these weren't just any scientists – they were like wizards of science, casting spells with chemicals and potions! Their mission was to find a magical potion that could make creatures live longer, just like in fairy tales.

Now, imagine their excitement when they stumbled upon something amazing – a pathway called sirtuin, which had the power to slow down aging in yeast cells! It was like discovering the secret ingredient to eternal youth!

But here's where it gets even more exciting: they wanted to see if they could use this magic on bigger creatures, like mice and maybe even humans. But how do you do that without turning them into mutant monsters?

That's where our hero, Konrad Howitz, comes in! Armed with his trusty chemistry set, he went on a quest to find chemicals that could boost sirtuin activity without any funny business.

And guess what? He struck gold – or should I say, he struck

strawberries and Chinese lacquer trees! Yep, he found magical compounds hidden in plants that could supercharge sirtuins and make them work ten times faster!

Now, these compounds weren't just any old chemicals – they were like rare gems, so precious that most scientists couldn't believe their luck! But our brave researchers knew they had found something special.

You see, these compounds were like keys that unlocked the body's natural defenses, making creatures healthier, stronger, and able to fight off diseases. They were like the superheroes of science, protecting us from the evils of aging!

And so, the quest continues as scientists work tirelessly to harness the power of these magical compounds and unlock the secrets of eternal youth. Who knows? Maybe one day, we'll all be sipping strawberry smoothies and living happily ever after, thanks to the magic of science!

Source: Adapted from David A. Sinclair PhD

Antioxidants

A lright, let's talk about antioxidants – those little superheroes that were all the rage back in 2002!

So, picture this: antioxidants are like the guardians of your body, fighting off the bad guys called free radicals that can damage your cells and make you age faster. One of these superhero antioxidants is called resveratrol, and it's found in red wine and some plants when they're feeling stressed out.

Now, some folks thought resveratrol might be the secret behind the "French paradox" – you know, how the French can eat all that yummy cheese and butter but still have healthier hearts than others? It's like they had a secret weapon hiding in their wine glasses!

Turns out, resveratrol was even more powerful than scientists thought! When they tested it in the lab, it outperformed other antioxidants like **Fisetin** and **Butein**. It was like the MVP of the antioxidant team!

To study how these antioxidants work, scientists looked at yeast cells – those tiny guys that multiply until they kick the bucket. And guess what? The less you disturb them (like putting them in the fridge for a nap), the longer they live! It's like they're living their best life on the dining room table, just chilling and multiplying.

So, next time you raise a glass of red wine, remember you're not just sipping on a tasty beverage – you're also getting a dose of antioxidants that could help keep you feeling young and spry! Cheers to that!

◆ ◆ ◆

Resveratrol in Grapes

Once upon a time, scientists got excited about something called resveratrol – it's like a magic potion found in grapes and red wine that might help us live longer!

So, they decided to do some experiments, starting with tiny creatures called yeast. They gave these yeast cells some resveratrol, and guess what? They lived longer! It's like they discovered the fountain of youth in a bottle of red wine!

But the fun didn't stop there. They tried resveratrol on fruit flies, and those little guys lived longer too! It was like they were getting extra years of life just by munching on some yeast paste mixed with resveratrol.

Then, they tried it on worms, and even human cells in dishes. And guess what? They all became stronger and more resistant to damage! It's like resveratrol was turning them into little superheroes!

But the real surprise came when they tested it on chubby mice. At first, the mice stayed chubby, and one scientist thought the experiment was a big flop. But when they checked inside those chubby mice, they found something amazing – their insides were super healthy! It's like resveratrol was keeping their hearts, livers, and muscles in tip-top shape!

And get this – some of those mice lived 20 percent longer than normal mice! That's like adding years to your life just by sipping on some red wine – talk about a happy hour!

But here's the kicker – even though resveratrol was like a superhero for animals, it wasn't very good at helping humans. It wasn't strong enough and didn't dissolve well in our tummies. But hey, it was a good start!

So, while we might not be able to guzzle down hundreds of glasses of red wine every day to stay young, at least we know that scientists are working hard to find ways to keep us feeling spry and youthful. Cheers to that – with a glass of grape juice, of course!

Once upon a time, scientists discovered something super cool – they found out that a chemical called **resveratrol**, found in grapes and red wine, can activate special proteins in our bodies called sirtuins. These sirtuins help us stay healthy and live longer!

But wait, there's more! Scientists didn't stop there. They went on a hunt for other chemicals that could do an even better job at activating these sirtuins. They called these compounds STACs, which stands for Sirtuin-Activating Compounds. It's like they were searching for a superhero team to keep us feeling young and strong!

And guess what? They found some awesome STACs like SRT 1720 and SR2104. When they gave these to mice, the mice stayed healthier for longer, just like they found the fountain of youth in a bottle!

But the fun doesn't end there. Another superhero in this story is NAD, which stands for Nicotinamide Adenine Dinucleotide. It's like the fuel that powers our sirtuin superheroes. Without enough NAD, our sirtuins can't do their job properly, and we might start feeling not-so-great.

Scientists found out that as we get older, our NAD levels drop, which isn't great news for our sirtuins. So, they started looking for ways to boost NAD levels, starting with experiments on yeast cells. Luckily, yeast cells are like little superheroes themselves, so even if things didn't go perfectly, it wasn't a big deal.

Now, scientists are trying to figure out if they can boost NAD levels safely in humans. It's like they're trying to find a way to give our bodies a little extra power-up to keep us feeling our best for as long as possible!

So, thanks to the curious minds of scientists, we might just have a secret weapon to help us stay healthy and strong for years to come. Who knew that grapes, red wine, and yeast could hold the key to unlocking the secrets of a longer, healthier life? Science sure is amazing!

◆ ◆ ◆

Magic Potion- Vitamins B3.

Once upon a time, a scientist named Charles Brenner found something super exciting while tinkering in his lab. He discovered that a special type of vitamin B3 called nicotinamide riboside, or NR for short, can help make our bodies stronger and healthier.

It's like a magic potion, but instead of dragons and wizards, we're talking about vitamins and science!

NR is like a superhero for our bodies. It boosts a special fuel called **NAD**, which helps keep our cells healthy and happy. Imagine NAD as the gas that makes our bodies run smoothly, and NR is the key ingredient that helps make more of this important fuel.

But wait, there's more! Another cool chemical called **nicotinamide mononucleotide, or NMN**, does a similar job. It's like NR's sidekick, helping to boost NAD levels and make our bodies feel like they're in tip-top shape.

When scientists gave NMN to mice, something amazing happened – the mice became super mice! They could run faster, jump higher, and even remember things better. It's like they became superheroes in their own little mouse world.

But here's the best part: these special chemicals aren't just for mice. Scientists are testing them in humans too, to see if they can make us healthier and stronger too. So far, it looks promising, and people are really excited about the possibilities!

So, thanks to the curious minds of scientists like Charles Brenner and his team, we might just have a secret weapon to help us stay healthier and happier for years to come. Who knew that vitamins and chemicals could hold the key to unlocking our body's superpowers? Science is truly amazing!

Alright, kiddos, let's dive into the fascinating world of how our bodies stay young and healthy! Imagine our cells as tiny superheroes fighting off the bad guys of aging. Now, picture this: there are special molecules called **NAD boosters** that help these superheroes stay strong and keep our bodies running smoothly.

These NAD boosters, like NMN (nicotinamide mononucleotide), are like magic potions that make our cells feel supercharged! They help our cells repair damage and keep our bodies feeling young and vibrant. It's like giving our bodies a big, superhero-sized hug from the inside.

But how do these NAD boosters work their magic? Well,

scientists are still figuring out all the details, but here's the gist of it: they help keep our cells' instruction manual, called the epigenome, nice and tidy. It's like tidying up your room to keep things organized and running smoothly.

When our cells get a boost from these NAD boosters, they become super efficient at fixing any damage that pops up. It's like giving our superhero cells an extra power-up so they can keep fighting off the bad guys of aging.

And get this: these NAD boosters might even be able to turn back the clock a bit! Imagine reversing aging in some parts of your body – it's like a real-life superhero movie right in our cells!

But wait, there's more! These NAD boosters could also help keep our eggs in tip-top shape. For lady mice, this means their eggs stay healthy and strong, just like in young mice. And for human ladies, it could mean keeping their baby-making parts healthy and happy too!

So, thanks to these amazing NAD boosters, our cells can stay young, strong, and ready to take on whatever life throws their way. It's like having a whole team of superheroes inside us, fighting off aging and keeping us feeling awesome!

Journey into the Future!

Wow, kids, buckle up for a mind-blowing journey into the future! Imagine if I told you that scientists are like detectives on a mission to crack the code of aging and extend our lives way beyond what we ever thought possible. How cool is that?

Picture this: there are these incredible molecules – let's call them superhero potions – that scientists are cooking up in their labs. These potions might just be the key to unlocking longer, healthier lives for all of us. But here's the kicker: there are so many different potions to choose from! It's like being in a candy store with endless options.

Maybe one potion, let's call it the AMPK activator, could give us an extra five years of supercharged health. Or perhaps a combination of potions, along with things like eating your veggies and doing jumping jacks, could give us a whopping couple of decades more!

But guess what? Just when you think things can't get any cooler, there's a plot twist! Scientists are uncovering new potions every day – some from sneaky little microorganisms, others from beautiful flowers – that could be even more powerful than anything we've seen before!

And get this: these potions aren't just sitting on shelves gathering dust. Nope, they're racing through human trials, getting ready to rock our world with their epic anti-aging powers. It's like watching superheroes gearing up for the ultimate battle against aging!

But here's the best part: even if none of these potions existed (which is hard to imagine because they're so darn cool), we'd still be on track for longer, healthier lives. That's right – scientists, engineers, and all sorts of clever folks are teaming up to revolutionize the way we live, and it's going to be epic!

So, kids, hold onto your hats because the future is looking bright, and who knows? Maybe one day, we'll all be living longer, healthier lives thanks to a few little potions and a whole lot of brainpower. How's that for a surprise twist in the story of humanity?

REFERENCES

1. Anderson, R. N. (2000). The Ten Leading Causes of Death in the U.S.: Final Data for 2000.

2. Ries, L. A. et al. (1999). SEER Cancer Statistics Review, 1973-1999. National Cancer Institute, Bethesda, Maryland. Late-stage diagnosis of lung, cervical, and ovarian cancers due to lack of screening.

3. Recommended Dietary Allowances (RDA). (1989). Last revised in 1989. Not optimized and do not accommodate unusual requirements from diseases or environmental stress.

4. Vermeulen, E. G. et al. (2000). Effect of Homocysteine-Lowering Treatment with Folic Acid Plus Vitamin B6 on Progression of Subclinical Atherosclerosis. The Lancet, 355(9203), 517-522.

5. Gilbert, S. F. The Genetic Core of Development: Differential Gene Expression. A gene both directs and is directed at protein synthesis.

6. Antisense RNA-mediated inhibition of MMR-

I expression reduces invasiveness in human chondrosarcoma. Journal of Orthopaedic Research, Nov; 21(6), 1063-1070.

7. RNA interference-mediated silencing of the S100A10 gene diminishes plasmin generation and invasiveness in Colo 222 colorectal cancer cells. Cellular and Molecular Biology, Oct 21. [Online]. Available: https://www.genscript.com/site2/document/897_20050927212002.PDF

8. Bressler, N. M. et al. (2003). Potential Public Health Impact of Age-Related Eye Disease Study Results: AREDS Report #11. Archives of Ophthalmology, Nov; 21(11), 1621-1624.

9. For information on the nutritional treatment of Autistic disorders, refer to the DAN (Defeat Autism Now) Protocols at the Autism Research Institute. [Online]. Available: www.autism.com/ari/contents.html

10. Frassetto, L. A. et al. (1998). Benefits from an Alkalizing Diet. The American Journal of Clinical Nutrition, 68, 516-523.

11. Shuster, J. et al. (1992). Soft Drink Consumption and Urinary Stone Recurrence. Journal of Clinical Epidemiology, Aug; 45(8), 911-916. Stone formation is

attributed to phosphoric acid in soft drinks (regular or diet).

12. Fernando, G. R., Martha, R. M., & Evangelina. (1998). Consumption of Soft Drinks with Phosphoric Acid as a Risk Factor for the Development of Hypocalcemia in Postmenopausal Women. Journal of Clinical Epidemiology, Oct; 52(10), 1007-1010.

13. Ferguson, A. (1995). Lactase deficiency occurs in 50-90% of most populations depending on race, with white Western Europeans being an exception. Mechanisms in Adverse Reactions to Food. Allergy, 50, 32-38.

14. Insoluble fiber increases stool size and shortens transit time in the intestine, potentially inhibiting the metabolism of carcinogens and the proliferation of "bad" bacteria. Dietary Fiber in Food and Protection Against Colorectal Cancer. The Lancet, May 3, 361(9368), 1426-1501.

15. Crawford, M. et al. (2000). The Role of Plant-Derived Omega-3 Fatty Acids in Human Nutrition. The American Journal of Nutrition Methods, 44(5-6), 263-265.

16. Kris-Etherton, P. M. et al. (2002). Fish Consumption, Fish Oil, Omega-3 Fatty Acids, and Cardiovascular Disease. Circulation, Nov 19; 106(21), 2747-2757.

17. Moncada, S. and Hinges, A. (1993). The L-Arginine-Nitric-Oxide Pathway. The New England Journal of Medicine, Dec 30; 329(27), 2002-2012.

18. Wang, B. Y. et al. (1999). Regression of Atherosclerosis: Role of Nitric Oxide and Apoptosis. Circulation, 99, 1236-1241.

19. Toxoplasma gondii is a parasite prevalent in wild and domestic animals worldwide. Up to 3 million women in the U.S. have acquired sexually transmitted Trichomonas vaginalis. National Institute of Allergy and Infectious Diseases (NIAID), Nov 1, 1993.

20. Lambert, J. D. and Yang, C. S. (2003). Green Tea, Rich in Catechin, a Polyphenol Antioxidant; Catechins in Black Tea Lost During Processing. Green Tea Exhibits Cancer Chemopreventive Activity.

21. Sierksma, A. et al. (2002). Moderate Alcohol Consumption Reduces Plasma C-reactive Protein and

Fibrinogen Levels. European Journal of Clinical Nutrition, Nov, 56(11), 1130-1136.

22. Acrylamide, a Potent Carcinogen, Found in Baked and Fried Carbohydrate-Rich Foods such as French Fries, Potato Chips, and Other Snack Foods, Alongside High Levels of Unhealthy Fat, Starch, Sugar, and Salt. (World Health Organization, 2002. Consultation on the Health Implications of Acrylamide in Food. Geneva, 25-27 June 2002).

23. After Adjustment for Established Risk Factors, Each Increment of 1 in BMI Increases the Risk of Heart Failure by 5% for Men and 7% for Women. (Kanchaniya, S. et al. 2002. "Obesity and the Risk of Heart Failure". New England Journal of Medicine, Aug 1; 347(5), 305-313).

24. In 2000, 64.5% of Americans Were Overweight, and 30.5% Were Obese. (Flegel, K. M. et al. 2002. "Obesity", TMA. Oct 9; 238(4), 1723-1727).

25. Insulin Resistance Plays a Role in the Pathogenesis of Alzheimer's Disease. (Watson, G. S. and Craft, S. 2003. "The Role of Insulin Resistance in the Pathogenesis of Alzheimer's Disease". CNS Drugs, 17(1), 27-45).

26. Insulin Resistance is Linked to Coronary Artery Disease. (Grant, P. J. 2003. "The Genetics of Atherothrombotic Disorders", Journal of Thrombosis and Hemostasis, Jul; 1(5), 1381-1390).

27. Nonalcoholic Steatohepatitis (Fatty Liver) is Associated with Insulin Resistance. (Scherer, A. and Lu, Y. C. 2003. "Nonalcoholic Steatohepatitis and Insulin Resistance", Acta Clinica Belgica, Mar-Apr; 58(2), 81-91).

28. Insulin Resistance Predicts Age-Related Diseases. (Facchini, S. et al. 2001. "Insulin Resistance as a Predictor of Age-Related Diseases", Journal of Clinical Endocrinology and Metabolism, Aug; 86(8), 3574-3528).

29. Oral Administration of Rac-Alpha-Lipoic-Acid Modulates Insulin Sensitivity in Patients with Type 2 Diabetes Mellitus. (Jacob, S. et al. 1999. "Oral Administration of Rac-Alpha-Lipoic-Acid Modulates Insulin Sensitivity in Patients with Type 2 Diabetes Mellitus", Free Radical Biology & Medicine, 27(3-4), 304-314).

30. Storlien, L. H. et al. (1987). Fish Oil Prevents Insulin Resistance Induced by High-Fat Feeding in Rats. Science, 237(4817), 885-888.

31. Dean, W. (Year Not Provided). Metformin: The Most Effective Life Extension Drug Is Also a Safe, Effective Weight Loss Drug.

32. Barrett-Connor, E. L. (1995). Testosterone and Risk Factors for Cardiovascular Disease in Men. Diabetes & Metabolism, 21, 156-161.

33. Lifetime Risk of Developing Breast Cancer in Women with a Positive Test for the BRCA1 Mutation is Estimated at 80%, compared to 10% for Non-Carriers. (Lancaster, J. M. 1997).

34. BRCA1 and 2: A Genetic Link to Familial Breast and Ovarian Cancer. (Medscape Women's Health, Feb; 2(2), 7. Other Studies Cite a 92% Total Lifetime Risk).

35. Hang, et al. (2003). Striking Higher Frequency in Centenarians and Twins of mtDNA Mutation Causing Remodeling of Replication Origin in Leukocytes. Proceedings of the National Academy of Sciences of the United States of America, Feb 4; 100(3), 1116-1121.

36. Single-nucleotide polymorphisms (SNPs) involve a mutation of only a single nucleotide (A, T, C, G), resulting in multiple shapes.

37. Myers, R. H. et al. (1996). Apolipoprotein E Epsilon 4 Association with Dementia in a Population-Based Study: The Framingham Study. Neurology, Mar; 46(3), 673-677.

38. Pyles, R. B. (2001). The Association of Herpes Simplex Virus and Alzheimer's Disease: A Potential Synthesis of Genetic and Environmental Factors. Herpes, Nov; 8(3), 64-68.

39. Itzhaki, R. F. et al. (1997). Herpes Simplex Type 1 in Brain and Risk of Alzheimer's Disease. The Lancet, Jan 25; 349(9047), 241-244.

40. The Best Vegetarian Source of Preformed EPA is Wakame, a Type of Seaweed Containing 186 mg of EPA per 100 grams. Flaxseed Oil, Containing 650 mg of EPA, Meets the Minimum Daily Requirement, Advised Since Over 12 oz of Wakame Would Be Needed.

41. Hujoel, P. et al. (2000). Periodontal Disease and Coronary Artery Disease Risk. JAMA, Sept 20; 284(11), 1406-1410.

42. NSAIDs demonstrate a reduced incidence of diseases associated with stent inflammation. Individuals who regularly took NSAIDs for a minimum of two

years experienced a 60% decrease in Alzheimer's disease risk. (McLendon, B. M. et al. 2000. "Current and Future Treatments for Cognitive Deficits in Dementia", Current Psychiatry Reports, Feb; 2(1), 20-23).

43. Ridker, P. M. et al. (1998). C-Reactive Protein Adds to the Predictive Value of Total and HDL Cholesterol in Determining Risk of First Myocardial Infarction. Circulation, May 26; 97(20), 2007-2011.

44. Patients with IL-1B 31C→T mutation face an increased risk of heightened inflammation from a given stimulus compared to those without this genetic variation. Inhibitory effects against inflammation resulting from this genetic defect can be observed with Fish Oil (EPA/DHA), Milk Thistle, Curcumin, Boswellia, and Licorice. (Sueoka, N. et al. 2001. "A New Function of Green Tea: Prevention of Lifestyle-Related Diseases", Annals of the New York Academy of Sciences, Apr; 928, 274-280).

45. Yokota, J. et al. (2003). Genetic Alterations Responsible for Metastatic Phenotypes of Lung Cancer Cells. Clinical & Experimental Metastasis, 20(3), 189-193. One gene associated with lung cancer is inactivated

in 50% of Lung Cancers by deletions, mutations, and methylation. (Alpha 1 Antitrypsin Deficiency predisposes individuals to early emphysema, particularly if they smoke).

46. Wolff, M. S., and Toniolo, P. G. (1995). Environmental Organochlorine (Pesticides) Exposure: A Potential Etiologic Factor in Breast Cancer. Environmental Health Perspectives, Oct; 103(Supp7), 141-145.

47. Husemoen, L. L. et al. (2004). Effect of Lifestyle Factors on Plasma Total Homocysteine Concentrations in Relation to MTHFR (C677T) Genotype. European Clinical Nutrition, March 31, 2004.

48. Coal-fired power plants are the largest industrial emitters of Mercury, producing over one-third of all mercury pollution in the U.S.

49. Song, N. et al. (2001). CYP-IA1 Polymorphism and Risk of Lung Cancer in Relation to Smoking. Carcinogenesis, Jan; 22(1), 11-16.

50. Payame, H. et al. (2001). Parkinson's Disease, CYP2D6 Polymorphism, and Age. Neurology, May 22; 56(40), 1363-1370.

51. Sheng, H. et al. (2004). Tacrolimus Dosing in Adult Lung Transplant Patients Is Related to Cytochrome P450 345 Gene Polymorphism. Journal of Clinical Pharmacology, Feb; 44(2), 135-140.

52. Konishi, et al. (2003). The HDH 3#2 and CYP 2Ei C2 Alleles Increase the Risk of Alcoholism in Mexican American Men. Experimental Molecular Pathology, Apr; 74(2), 183-189. For the DR-70 test, visit the AMDL, Inc. website: www.amde.com/products/dr-70/index.html.

53. Up to 48% of Stomach Cancer cases may be attributed to the GSTM I Null polymorphism combined with mutations of the IL-1B and NAT1 genes. (Gonzalez, C. A. et al. 2002. "Genetic Susceptibility and Gastric Cancer Risk", International Journal of Cancer, Jul 20; 100(3), 249-260).

54. Wang, Y. C. and Bachrach, U. (2002). The Specific Anticancer Activity of Green Tea Epigallocatechin-3-Gallate (EGCG). Amino Acids, 22(2) 131-143.

55. Polich, J. (2001). Cognitive Effects of a Ginkgo Biloba/Vinpocetine Compound in Normal Adults: Systematic Assessment of Perception, Attention,

and Memory. Human Psychopharmacology, Jul; 16(5), 409-416.

56. Sharman, E. H. et al. (2002). Reversal of Biochemical and Behavioral Parameters of Brain Aging by Melatonin and Acetyl L-Carnitine. Brain Research, Dec 13; 957(2), 223-230.

57. Chung, S. Y. et al. (1995). Administration of Phosphatidylcholine Increases Acetylcholine Concentration and Improves Memory in Dementia Mice. Journal of Nutrition, Jun; 125(6), 1484-1489.

58. Hackbert, L. and Heiman, J. R. (2002). Mental and Physical Sexual Arousal Ratings Increase Significantly in Response to an Acute Dose of DHEA in Postmenopausal Women.

59. Blackman, M. R. et al. (2002). Most Anti-Aging Physicians Recommend Low-Dose High-Frequency Growth Hormone Injection Protocols to Reduce Side Effects.

60. Touitou, Y. (2001). Human Aging and Melatonin: Clinical Relevance. Experimental Gerontology, Jul; 36(7), 1083-1100.

61. Fraschini, E. et al. (1998). Melatonin Involvement in Immunity and Cancer. Biological Signals, Jan; 7(1), 61-72.

62. Church, T. S. et al. (2003). Reduction of C-Reactive Protein Levels through the Use of a Multivitamin. The American Journal of Medicine, Dec 15; 115(9), 702-707.

63. Ames, B. N., Elson-Schwab, I., & Silver, E. A. (2002). High-Dose Vitamin Therapy Stimulates Variant Enzymes with Decreased Coenzyme Binding Affinity: Relevance to Genetic Diseases and Polymorphisms. The American Journal of Nutrition, Apr; 75(4), 616-658.

GINKGO BILOBA

WORD LIST: TRICKY BOOK TERMS

Here's a list of all the fancy or tricky words you might come across in this book called 'Live longer, Love longer, age is a treatable disease'."

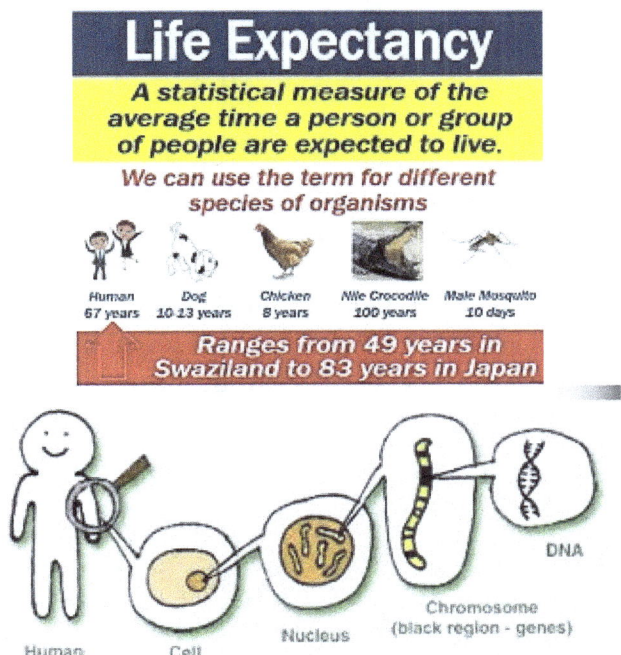

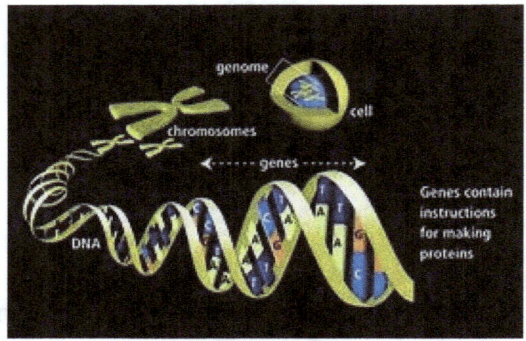

"ALLELE: Think of alleles as different flavors of a gene. Each one brings its unique twist to the genetic recipe. For instance, a 'bad news allele' is like a gene version that leads to trouble, like causing diseases.

AMINO ACID: Picture amino acids as the Lego blocks of proteins. When your body's making proteins, it's like stringing together different Lego pieces to build something cool.

ANTAGONISTIC PLEIOTROPY: Imagine this theory as Mother Nature's quirky way of explaining why we age. It's like saying that a gene that shortens our lifespan in old age might stick around if its early benefits outweigh the later drawbacks. It's like getting a free pass early on, but then dealing with the consequences later."

"BASE: Think of these as the alphabet letters in the Book of Life, but instead of ABCs, it's more like ATCGs. Each letter—A, C, T, and G—is a chemical buddy that helps spell out our genetic story. And just to keep things spicy, RNA swaps out T for U.

BASE PAIR: Picture DNA as a zipper with little teeth. These teeth are the bases, and they love to cuddle up with their perfect match on the other side. It's like a dance where C always pairs with G, and A has a crush on T (or U, if you're talking RNA).

BIOTRACKING / BIOHACKING: Imagine playing detective with your body, using gadgets and tests to snoop around and figure out what makes it tick. It's like being your health detective, but don't mix it up with biohacking, where people DIY their way to superhuman upgrades."

"CANCER: Picture cells throwing a wild party where they forget to stop growing. They form clumps called tumors and start gate-crashing into other parts of the body in a process called metastasis.

CELL: Imagine cells as the tiny building blocks of life. They're like Legos that make everything from a single-celled yeast to a giant blue whale. And inside each cell, there's a whole bunch of important stuff like proteins, fats, carbs, and DNA, keeping things running smoothly.

CELLULAR REPROGRAMMING: It's like hitting the rewind button on cells, taking them back to an earlier stage of development.

CELLULAR SENESCENCE: Think of this as cells hitting the snooze button on dividing and instead causing trouble by releasing grumpy molecules. Even though they're like zombie cells, they're still kicking around and causing havoc with their inflammatory chatter.

CHROMATIN: Imagine DNA cozying up with proteins like a bunch of spaghetti wrapped around meatballs. When it's all spread out, genes can do their thing (that's euchromatin), but when it's all packed up tight, it's like gene silence mode (heterochromatin).

CHROMOSOME: It's like DNA packing itself up into tidy little bundles held together with protein glue. Different organisms have different numbers of these bundles—humans rock 23 pairs.

COMPLEMENTARY: Think of DNA and RNA sequences as dance partners who fit together perfectly like puzzle pieces. Each letter in one sequence snuggles up with its perfect match in the other sequence, like a genetic love story.

CRISPR: Imagine bacteria and archaea with a secret weapon—a genetic scissors called CRISPR. It's like a DNA ninja that cuts out specific bits of genetic code, making precision edits like a molecular surgeon."

"DAF-16/FOXO: Imagine DAF-16/FOXO as the superhero buddy of Sirtuins, a genetic control freak who activates the body's defense squad. This activation of defense genes can stretch out the lifespan of all sorts of critters, from tiny worms to maybe even us humans. It's like the VIP pass for a longer life in the animal kingdom.

DEACETYLATION: Picture tiny enzymes playing tag with protein tags. When they tag 'em with acetyls, it's like turning on a gene's 'off' switch. But then along comes sirtuins, the ultimate tag remover, wiping away those acetyls like they're cleaning a slate. It's like a genetic cleanup crew getting rid of all the unnecessary clutter.

DEMETHYLATION: It's like a molecular makeover where enzymes sweep in and remove methyl tags, those pesky little markers that mess with gene expression. And just like that, the gene's back to its original, fresh-faced self.

DISPOSABLE SOMA: Think of this as nature's big dilemma: do you go all out on fast reproduction or invest in a long-lasting body? It's like trying to decide whether to splurge on a flashy car or save up for a sturdy house. Because in the wild, resources are tight, and you can't have it all.

DNA: It's the instruction manual for life, shaped like a twisted ladder made of a molecule called deoxyribonucleic acid. And just like a zipper, it's got these letters—A, C, T, and G—lined up on each side, always sticking together in pairs like genetic BFFs.

DNA DOUBLE-STRAND BREAK (DSB): Imagine your DNA as a delicate necklace that accidentally snaps in two. But fear not, cells are like expert jewelry repairers, swooping in to mend the broken strands. Sometimes they even add a little bling, changing the DNA sequence in the process. It's like molecular DIY gone wild.

DNA METHYLATION CLOCK: Picture this as your body's biological clock, but instead of ticking seconds, it's marking time with tiny methyl tags on your DNA. These tags can predict your lifespan and even turn back the clock on aging, like a magical rewind button for your cells."

ENZYME: Think of enzymes as the super-speedy chefs of the cellular kitchen. They're like protein master chefs, whipping up chemical reactions at lightning speed that would otherwise take forever. Take sirtuins, for example—they're the cleanup crew using NAD to clear out acetyl tags from histones like they're scrubbing dirty dishes.

EPIGENETIC: Imagine your genes wearing little sticky notes that tell the cell what to do and when to do it, without actually changing the DNA code itself. It's like leaving a note in a book saying "Ignore this page," so even though the reader skips it, the book's still the same.

EPIGENETIC DRIFT AND EPIGENETIC NOISE: It's like your body's genetic playlist getting shuffled around as you age, thanks to changes in methylation—kinda like turning up the volume on some genes and muting others. And just like a noisy neighbor, environmental factors can crank up the volume even more, leading to epigenetic chaos.

EXDIFFERENTIATION: Picture cells losing their sense of self because of all the genetic noise, like forgetting who they are at a wild party. This identity crisis could be a major culprit behind the aging process, turning cells from sophisticated specialists into confused party crashers.

EXTRACHROMOSOMAL RIBOSOMAL DNA CIRCLE (ERC):

Imagine little troublemakers causing chaos in the cellular neighborhood by distracting the sirtuins and breaking up the nucleolus—the cell's headquarters for making proteins. It's like throwing a wild party that ends up trashing the house and causing premature aging."

GENE: Imagine genes as tiny recipe cards made of DNA. Each card holds instructions for building a specific molecular machine that helps cells, organisms, or even viruses do their thing.

GENE EXPRESSION: It's like a gene throwing a big party where it shows off its stuff. When a gene is "turned on," it's like the DJ blasting tunes and everyone dancing. This can happen when the gene's DNA is copied into RNA and then translated into a chain of amino acids, aka proteins. The more the gene's expression, the louder the party and the more protein it pumps out.

GENE THERAPY: Think of this as molecular medicine, where doctors deliver healthy DNA to fix up wonky cells. It's like giving a cell a genetic makeover by adding a shiny new DNA sequence to its genome. They use sneaky viruses as delivery drivers to drop off the new genes in the right place, like genetic superheroes swooping in to save the day.

GENETICALLY MODIFIED ORGANISM (GMO): It's like playing genetic mad scientist, where we tweak an organism's DNA using fancy tools. We can genetically engineer anything from microbes to plants to animals, giving them new superpowers or making them more resistant to pests and diseases.

GENOME: Picture this as the ultimate instruction manual for life, containing all the DNA sequences needed to build and run an organism. It's like a massive blueprint that tells cells what to do and how to do it.

GENOMICS: It's like detective work for DNA, where scientists dig deep into an organism's genetic code. They study everything from the DNA sequence itself to how genes are organized and controlled, as well as the molecules that interact with DNA, and how all these pieces affect cell growth and function.

GERM CELLS: Imagine these as the VIPs of reproduction—eggs, sperm, and the cells that make them. Any genetic tweaks made to these cells could be passed down to future generations, like inheriting your grandma's recipe for the perfect chocolate chip cookies. So, when we edit the DNA in these early cells, it's like making changes that will stick around in the family tree for generations to come.

HISTONES: Think of histones as the DNA's interior decorators—they're the reason why three feet of DNA can cozy up inside a tiny cell. Each DNA strand wraps around histones like a fancy scarf around a neck, forming a neat little package. And just like a good decorator, enzymes like sirtuins come in to add or remove chemical touches, controlling how tightly the DNA is packed. Tight packaging creates 'silent' areas where genes snooze, while looser packaging is like opening the curtains to let the genes dance.

HORMESIS: It's like the superhero mantra: what doesn't kill you makes you stronger. Imagine plants getting a little dose of herbicide—they don't keel over, they grow even faster. It's like giving your body a mini workout that actually boosts its repair processes, making cells stronger and healthier.

INFORMATION THEORY OF AGING: Picture aging as your brain's hard drive gradually losing files over time, especially the juicy epigenetic ones. But here's the twist: it's not all doom and gloom. Scientists reckon we can recover a lot of that lost info, like finding those missing cat videos buried in your computer's archives.

METFORMIN: Think of metformin as a fancy potion brewed from French hellebore, but instead of turning people into frogs, it helps treat type 2 diabetes. And hey, bonus—some folks think it might even be the fountain of youth in pill form!

MITOCHONDRIA: Meet the powerhouse of the cell—mitochondria! They're like the Energizer Bunny, breaking down snacks to make energy through a process called cellular respiration. And get this—they even have their little circular genome, like a tiny DNA club inside your cells.

MUTATION: It's like DNA playing a game of genetic Mad Libs, where a letter gets swapped out for another. Sometimes it's no biggie, just a fun twist in the genetic code that makes life interesting. But other times, it's like hitting the genetic jackpot for diseases. Mutations can happen naturally, like getting a sunburn from too much UV light, or even with a little help from enzymes playing genetic copycat. And hey, sometimes we humans even play genetic engineering and deliberately mix things up for fun (and science, of course!).

NAD: Meet NAD, the multitasking molecule that's busier than a bee at a flower convention. It's like the Swiss Army knife of chemicals, pulling off over 500 different chemical tricks. And when teamed up with sirtuins, it's like the dynamic duo of gene regulation, flipping switches to turn genes off or give them superhero cell-saving powers. Plus, it even gets a little extra kick with that "+" sign, showing it's ready to party without a hydrogen atom in tow.

NUCLEASE: Imagine nuclease as the cell's ultimate DNA and RNA ninja, slicing and dicing genetic material like a master chef. Breaking just one strand is like making a small cut while breaking both strands is like a genetic karate chop. Endonucleases strike in the middle of DNA or RNA, while exonucleases go for the ends, like genetic slicing experts. And when it comes to genome engineering, tools like Cas9 and I-Ppol are like the cool kids on the block, wielding their endonuclease powers with precision.

NUCLEIC ACIDS OR NUCLEOTIDES: Think of nucleic acids as the

building blocks of genetic Lego sets, with each piece made up of a base, a sugar, and a phosphate group. The sugars and phosphates link together to form the backbone, while the bases pair up like genetic soulmates, forming the famous A-T and C-G base pairs that hold DNA strands together.

NUCLEOLUS: Picture the nucleolus as the cell's cozy knitting circle, nestled inside the nucleus and buzzing with activity. It's like the nerve center for making proteins, where the genes for building ribosomes—the protein factories of the cell—are chilling out. It's where all the cool kids gather to stitch together amino acids and whip up protein masterpieces.

PATHOGEN: Meet the troublemakers of the microbe world —pathogens! They're like the party crashers who show up uninvited and wreak havoc, causing all sorts of illnesses. Now, most microbes are harmless little critters, but some strains or species are like the bad apples of the bunch, ready to stir up trouble whenever they get the chance.

PROTEIN: Think of proteins as the Swiss Army knives of the cell —they're like versatile tools folded into intricate shapes. Each protein has its special job, whether it's building stuff, breaking stuff down, or just keeping the cell running smoothly. And just like how we need a balanced diet, cells need proteins along with lipids, carbs, and nucleic acids to stay healthy and happy.

RAPAMYCIN: Say hello to rapamycin, also known as sirolimus —the multitasking marvel that's like the superhero of immune suppression in humans. It's like the bouncer at the cellular nightclub, putting the brakes on rowdy T cells and B cells by making them less sensitive to a key signaling molecule called interleukin-2. And get this—it even extends its lifespan by hitting the brakes on mTOR, the cellular growth regulator.

REDIFFERENTIATION: Picture this as the ultimate age-defying makeover for cells—it's like hitting the reset button on those pesky epigenetic changes that come with aging. It's like turning back the clock and giving cells a fresh start, reversing all the wrinkles and gray hairs of cellular aging.

RIBOSOMAL DNA (rDNA): Meet the VIPs of protein production—ribosomal DNA! They're like the architects behind every protein-making project in the cell, providing the genetic blueprint for ribosomal RNA, the building blocks of the ribosome—the cell's protein-making factory. It's like the DNA's way of saying, "Let's get this protein party started!"

RNA: Think of RNA as DNA's trusty sidekick, transcribed from the DNA blueprint and ready to kickstart the protein-making process. It's like the messenger delivering orders from the DNA headquarters to the protein production line. And when it comes to genome editing with CRISPR, RNA is like the GPS guiding the molecular scissors to their target spots in the DNA.

SENOLYTICS: Imagine senolytics as the superhero medicines of the future, on a mission to zap away those pesky senescent cells that cause trouble as we grow older. They're like the Avengers of anti-aging, fighting to slow down or even reverse all the aging-related issues.

SIRTUINS: Meet the guardians of longevity—sirtuins! They're like the secret agents of health, found in everything from tiny yeast to us humans. These cool enzymes need a special fuel called NAD* to work their magic. They're like cellular superheroes, swooping in to remove little tags from proteins, telling them to gear up and protect our cells from bad stuff like disease and death. And when we fast or exercise, our sirtuin, and NAD* levels go up, which is why these activities are like super healthy power-ups.

SOMATIC CELLS: Think of somatic cells as the team players in the body—they're all the cells except for the ones involved in making babies (eggs or sperm). Any changes or mutations they pick up won't be passed on to the next generation unless we're talking about cloning, which is a whole other story.

STEM CELLS: Imagine stem cells as the magical shape-shifters of the body—they're like superheroes with the power to turn into any type of cell or make more of themselves. Most cells in your body have already picked their career paths, but stem cells are like free agents, ready to jump in wherever they're needed. So, when your body gets damaged, these superhero cells swoop in to save the day and keep things running smoothly.

STRAND: Picture a strand as a fancy necklace made of tiny beads called nucleotides. When two strands of DNA come together, it's like they're holding hands, with their bases matching up to form pairs. And just like how DNA likes to cozy up in pairs, RNA prefers to go solo, like a lone wolf roaming around the cell.

SURVIVAL CIRCUIT: It's like the emergency backup system in cells, ready to kick into gear when things get tough. This ancient control system evolved to help cells focus on repair during hard times, like shifting energy away from growth and reproduction. But sometimes, the system gets stuck in survival mode, leading to all sorts of aging shenanigans. It's like the cell's way of saying, "Better safe than sorry!"

TELOMERES/TELOMERE LOSS: Imagine telomeres as the little caps at the end of our chromosomes, kinda like the plastic tips on your shoelaces or the burnt ends of a rope that keep it from fraying. But as we get older, these caps start to wear down, just like how your shoelaces get shorter with each tie. Eventually, when the telomeres get too short, it's like the cell hitting a roadblock—it can't divide anymore and just decides to take a nap, becoming what we call "senescent."

TRANSCRIPTION: Think of transcription as the genetic copycat game, where DNA instructions are copied onto a strand of RNA. It's like making a photocopy of a recipe so you can follow it in the kitchen. And just like how you need a special machine to make those copies, cells use an enzyme called RNA polymerase to do the job.

TRANSLATION: Picture translation as the protein-making dance party, where the instructions encoded in RNA are used to whip up proteins. It's like following a recipe to make a delicious cake, but instead of flour and eggs, cells use amino acid building blocks. And just like how a chef puts together ingredients to create a masterpiece, a molecular machine called the ribosome links these amino acids into a protein, folding it up into its final 3D shape.

VIRUS: So, imagine viruses as these sneaky little troublemakers who can't even bother to do their homework. Nope, instead, they crash at someone else's place, hijack their stuff, and throw a wild party where they replicate themselves like crazy. And get this—they're not even considered alive! They're like the freeloaders of the biological world, infecting everything from humans to plants to tiny microbes. But hey, at least our bodies have superhero immune systems to fend off these viral invaders, while bacteria and archaea have their own special CRISPR systems to kick viruses to the curb.

WADDINGTON'S LANDSCAPE: Picture this as the ultimate rollercoaster ride for cells during their fancy embryonic development phase. It's like a magical 3D map where stem cells start as little marbles and roll down into different valleys, each one leading to a unique path of cell destiny. It's like those choose-your-own-adventure books, but for cells!

XENOHORMESIS HYPOTHESIS: Here's a wild idea—our bodies are like secret agents, tuned in to the stress signals of other species, especially plants. It's like they have their plant radar to sense danger and gear up for battle. That's why we've got so many medicines coming from plants—they're like nature's way of giving us a heads-up when trouble's on the horizon.

Umar Syed RIP 1992-2016

Umar Syed Foundation Corporation is a dynamic private foundation based in Baytown, TX, dedicated to championing education, healthcare, and scientific advancement for underserved children in our community.

For inquiries, reach out to us via email at medicaltower@gmail.com or send mail to our office address:
5714 Comal Park Ct,
Houston, Texas 77059

Together, let's build a brighter future through education, health, and science for all.

Warm regards,
[Umar Syed Foundation Team]

www.ingramcontent.com/pod-product-compliance
Lightning Source LLC
Chambersburg PA
CBHW070420290526
45791CB00005B/1769

* 9 7 9 8 8 8 4 8 7 3 6 3 6 *